REVIEW FOR THE CLEP* GENERAL COLLEGE MATHEMATICS EXAMINATION

19th EDITION

By
Comex Staff

comex systems, inc.
5 Cold Hill Rd.
Suite 24
Mendham, NJ 07945

* "CLEP and College Level Examination Program are registered trademarks of The College Entrance Examination Board. These Materials have been prepared by Comex Systems, Inc., which bears sole responsibility for their contents."

Published by

comex systems, inc.
5 Cold Hill Rd., Suite 24
Mendham, NJ 07945

ISBN 1-56030-248-8

Table of Contents

CLEP* (College Level Examination Program)

WHAT IS CLEP?

CLEP is a nation-wide program of testing which began in 1965. Today thousands of colleges recognize CLEP as a way students may earn college credit. Each year hundreds of thousands of students take CLEP examinations. All examinations are designed and scored by the College Entrance Examination Board (CEEB). The purpose of each examination is to determine whether your current knowledge in a subject will qualify you for credit in that area at a particular college.

There are five general examinations. The subject areas are:

1. English Composition
2. Mathematics
3. Social Science / History
4. Natural Science
5. Humanities

Credits earned through these examinations replace basic liberal arts credits which are required by many colleges. Each of these general examinations is broad in coverage. Questions are from the wide range of subjects included in each of the major disciplines. Each examination covers a very wide topic of subjects; you are not expected to have in-depth knowledge of all areas.

HOW LONG ARE THE EXAMINATIONS?

Each CLEP General Examination is 1½ hours in length.

HOW MUCH DO THE EXAMINATIONS COST?

Currently, the fee to take each examination is $60.00. They may be taken one at a time or in any combination. (NOTE: Fees change periodically.)

You should expect to pay a non-refundable fee to the test center for administration costs. Each test center sets their own fee.

WHERE WILL THE EXAMINATIONS BE TAKEN?

The CEEB (College Entrance Examination Board) has designated certain schools in each state to serve as test centers for CLEP examinations. The same examinations are given at each test center. If you are a member of the armed forces, consult with the Education Services Officer at your base. Special testings are set up for military personnel.

You may obtain information regarding test locations and colleges that accept CLEP credits at www.collegeboard.com.

WHEN ARE THE TESTS GIVEN?

Check your test center for specific information. If you are serving with the United States Military, check with the Education Services Officer at your base to find out about the DANTES testing program. You will be given information about testing as applicable to military personnel.

HOW DO YOU REGISTER FOR AN EXAMINATION?

A registration form may be obtained from the test center where you plan to take the examination. Many centers require that you register (send registration form and fee for examinations to be taken) a month prior to your selected date.

WHEN WILL SCORES BE RECEIVED?

Most tests taken on the computer will be scored immediately (the English Composition with Essay must be graded first). If you are in the military and are taking the test on paper it will take up to 6 weeks to receive your scores. You may also request that a copy be sent to a college. The score you receive will be a scaled score. CEEB keeps a record of your scores on file for 20 years. You may obtain an additional copy or have a copy sent to a college if you contact:

College Board
PO BOX 6600
ATTN: Transcript Service
Princeton, NJ 08541

800-257-9558

IS IT NECESSARY TO BE ENROLLED IN A COLLEGE BEFORE YOU TAKE AN EXAMINATION?

Each college has established policies regarding CLEP. Check with the school you wish to attend. Many schools do not require enrollment before taking CLEP examinations.

HOW MANY CREDITS MAY BE EARNED?

Each college determines the number of credits that may be earned by CLEP examinations. Most colleges award six credits for achievement on a CLEP General Examination.

There is no correlation between the number of questions you answer correctly and the percentile level you achieve. The number will vary from test to test.

CAN THE SAME SCORES EARN A DIFFERENT NUMBER OF CREDITS AT DIFFERENT SCHOOLS?

Yes, different schools may require different levels of achievement. Your scores may earn more credits at one institution than at another. For example: if you achieve at the 25th percentile level, you could earn credit at a school which required the 25th percentile level; you could not earn credit at a school which required a higher level of achievement.

CAN CLEP CREDITS BE TRANSFERRED?

Yes, provided the school to which you transfer recognizes CLEP as a way to earn credit. Your scores will be evaluated according to the new school's policy.

CAN AN EXAMINATION BE RETAKEN?

Many schools allow you to retake an examination if you did not achieve the first time. Some do not. Check your particular school's policy before you retake an examination. Be realistic, if you almost achieved the level at which you could earn credit, do retake the examination. If your score was quite low, take the course it was designed to replace.

If you decide to retake an examination, six months must elapse before you do so. Scores on tests repeated earlier than six months will be canceled.

HOW TO USE THIS BOOK:

1. Complete the review material. Take the short tests included at the end of the lessons.
2. If you do well on the tests, continue. If you do not, review the explanatory information.
3. After completing the review material, take the practice examination at the back of the book. When you take this sample test, try to simulate the test situation as nearly as possible:
    a. Find a quiet spot where you will not be disturbed.
    b. Time yourself accurately.
    c. Practice using the coding system described in the next section.
4. Correct the tests. Determine weaknesses. Go back and review those areas in which you had difficulty.

THE CODING SYSTEM

On the CLEP exam, there is no penalty for wrong answers. Your score is computed based on the number of correct answers. When you are finished with the test make sure that every question is answered; however, you do have the ability to go back to unanswered questions, so do not feel pressured to answer each question in order. The CLEP tests allow you to flag the questions that you would like to return to. If you use the coding system detailed below, you can greatly increase your score.

Over the years, COMEX has developed a systematic approach to taking a multiple choice examination. This is called the coding system. It is designed to:

1. help you work through the examination as quickly as possible
2. have you answer the easy questions quickly
3. eliminate time wasted on difficult questions
4. help you eliminate incorrect answers and determine when to guess
5. have your questions prioritized for the second reading
6. use the recall factor. It is possible that a later question will trigger a thought that will help you answer a previous question. This can only be used in the section of the test you are currently working in.

How the Coding System Works

Managing your time on the exam can be as important as knowing the correct answers. If you spend too much time working on difficult questions which you have little or no knowledge about, you might not get as far as easy questions later that you would have gotten correct. This could cause a significant decrease in your score.

Below are some sample questions:

1. George Washington was:

   a. the father of King George Washington
   b. the father of Farah Washington
   c. the father of the Washington Laundry
   d. the father of Washington State
   e. the father of our country

As you read the questions, eliminate all the **wrong** answers:

a. father of King George Washington NO!
b. father of Farah Washington NO!
c. father of the Washington Laundry NO!
d. father of Washington State NO!
e. the father of our country YES. LEAVE IT ALONE.

The question now looks like this:

1. George Washington was:

~~a. the father of King George Washington~~
~~b. the father of Farah Washington~~
~~c. the father of the Washington Laundry~~
~~d. the father of Washington State~~
e. the father of our country

Click on the button next to the correct answer and click Next.

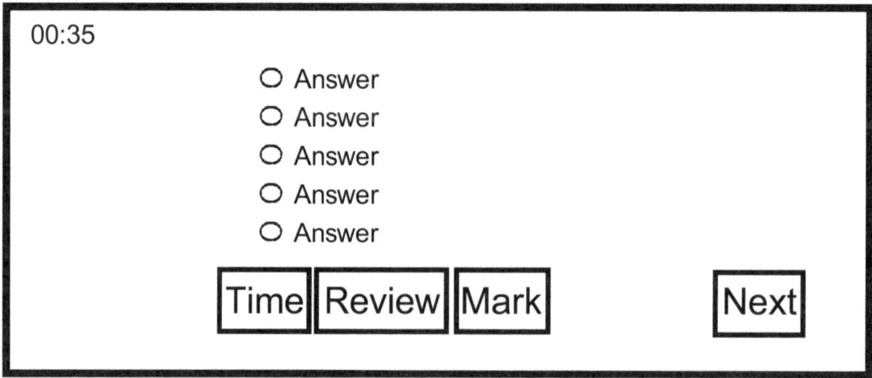

These are the buttons you must know how to use!

This question is now completed. When you get to the review process, this question will be sorted as answered. That will be your signal to not spend any more time with this question.

2. Abraham Lincoln was responsible for:

a. freeing the 495 freeway
b. freeing the slaves
c. freeing the Lincoln Memorial
d. freeing the south for industrialization
e. freeing the Potomac River

Go through the answers. Be sure to read ALL the choices.

a.	freeing the 495 freeway	NO!
b.	freeing the slaves	MAYBE
c.	freeing the Lincoln Memorial	NO!
d.	freeing the south for industrialization	MAYBE
e.	freeing the Potomac River	NO!

The question now looks like this:

2. Abraham Lincoln was responsible for:

~~a. freeing the 495 freeway~~
b. freeing the slaves
~~c. freeing the Lincoln Memorial~~
d. freeing the south for industrialization
~~e. freeing the Potomac River~~

Should you guess? You have very good odds of getting this question correct. Pick the choice that you feel is the best answer. Often your first guess will be the best. Before clicking the Next button, click on the Mark box. This will tell you later that you were able to eliminate 3 answers before guessing. Now click on Next to go on to the next question.

3. Franklin Roosevelt's greatest accomplishment was:

a. building the Panama Canal
b. solving the Great Depression
c. putting America to work
d. organizing the CCC Corps
e. instituting the income tax

Go through the answers:

a.	building the Panama Canal	NO! (different Roosevelt)
b.	solving the Great Depression	MAYBE
c.	putting America to work	MAYBE
d.	organizing the CCC Corps	MAYBE
e.	instituting the income tax	MAYBE

The question now looks like this:

3. Franklin Roosevelt's greatest accomplishment was:

    ~~a. building the Panama Canal~~
    b. solving the Great Depression
    c. putting America to work
    d. organizing the CCC Corps
    e. instituting the income tax

Should you answer this question now? Not yet. There might be a question later that contains information that would help you eliminate more of the answers. When you can only eliminate one answer, or none at all, your best course of action is to simply click on [Next]. This will bring up the next question and leave the status of that question as "unanswered" on the review list.

Now look at another question:

4. Casper P. Phudd III was noted for:

    a. rowing a boat
    b. sailing a boat
    c. building a boat
    d. designing a boat
    e. navigating a boat

Even if you have no idea of who Casper P. Phudd III is, read the answers:

a.	rowing a boat	I do not know.
b.	sailing a boat	I do not know.
c.	building a boat	I do not know.
d.	designing a boat	I do not know.
e.	navigating a boat	I do not know.

Since you cannot eliminate any of the answers, simply go on to the next question.

Try another question:

5. Clarence Q. Jerkwater III

    a. sailed the Atlantic Ocean
    b. drained the Atlantic Ocean
    c. flew over the Atlantic Ocean
    d. colored the Atlantic Ocean orange
    e. swam in the Atlantic Ocean

Even though you know nothing of Clarence Q. Jerkwater III, you read the answers.

a.	sailed the Atlantic Ocean	Possible.
b.	drained the Atlantic Ocean	No way!
c.	flew over the Atlantic Ocean	Maybe.
d.	colored the Atlantic Ocean orange	No way!
e.	swam in the Atlantic Ocean	Maybe.

The question now looks like this:

5. Clarence Q. Jerkwater III
   - a. sailed the Atlantic Ocean
   - ~~b. drained the Atlantic Ocean~~
   - c. flew over the Atlantic Ocean
   - ~~d. colored the Atlantic Ocean orange~~
   - e. swam in the Atlantic Ocean

Do you take a guess? Not on the first reading of the answers. Let us wait to see if the recall factor will help. Do not click on an answer, but do click Mark because two answers were eliminated. Click Next to get the next question.

Continue in this manner until you finish all the questions in the section. By working in this manner you have organized the questions to maximize your efficiency. When you finish with the last question click Review. This brings up the listing of all the questions. They will be listed in numerical order, with their status next to each question. This is not the way you want to view them; you want to view the questions sorted by status. Click on Status. Now the questions are sorted for you.

Now you are ready to go through the test a second time. Below is a review of what each status means, and tips on how to proceed through the review portion of the coding system.

- ♦ not answered with a check mark
  REVIEW ALL OF THESE QUESTIONS FIRST.

  You have eliminated two answers. Re-read them, trying to eliminate an additional answer. If you are able to do that, make your best guess between the remaining two answers but leave the question as "marked". You can re-review them after all questions have been answered.

  If you feel certain that you have the correct answer, remove the "mark", so that you do not spend any more time on that question.

- not answered without a check mark (status: unanswered).
  REVIEW ALL OF THESE QUESTIONS NEXT.

  You could only eliminate one answer. Re-read the questions, trying to eliminate additional answers. As above, if you can make a guess, do so, but leave the "mark" on the question.

  If you feel certain that you have chosen the correct answer, remove the "mark" so that you do not spend any more time on that question.

- answered with a check mark
  REVIEW THESE QUESTIONS AFTER ALL ABOVE QUESTIONS ARE ANSWERED

  You eliminated three answers and picked one of the remaining two. This question is answered, but could use some review. Do not review these questions until the remaining questions are answered.

- answered without a check mark (status: answered)
  DO NOT REVIEW THESE QUESTIONS.

  You knew the correct answer. Do not return to these questions.

Time management is also very important when taking the CLEP examinations. Be sure to click on the box that will display the time for you as you work your way through the test.

When there are 5 minutes left, make sure that you have every question answered. Be sure that you have guessed at the questions you were not sure about; you will get a better score than if the questions were left blank. Every correct answer will increase your score, and incorrect answers will not harm you.

All the questions are answered. Does this mean it is time to stop? Not if you want to get your highest score. All of the questions on which you took educated guesses have a check mark. Keep working on those problems, using all the allotted time. Do not waste time looking at any questions that do not have a check. You knew the correct answer and are done with them.

By using the coding system you will move quickly through the test and make sure that you see every question. It also allows you to concentrate your efforts on your strongest areas.

You can practice the system while you do your exercises and tests through the book with a piece of scratch paper. Put an "A" next to questions as you answer them. Put a check mark next to a question to refer back to it if you are not completely certain of the answer. Then use the system to go back through the test. The system is easy to master and will be an invaluable tool for all your test-taking.

You have now completed the portion of this book which is designed to improve your test-taking ability. When you do the practice exercises and take the sample test, use the techniques you have just learned.

You can use the coding system on any multiple choice exam. This will not only increase your score on that test, but it will also make you more comfortable with using the system. It has been demonstrated many times that the more comfortable you are when you are taking a test, the higher your score will be.

Tips for Testing Day

1. Get to the test center early. If you are taking the examination at a new location - check out how to get there **before** the day of the test.
2. If you do not understand the proctor's directions, ask questions before the test begins. You do not want to waste valuable time clarifying instructions.
3. Do not quit. Keep going over questions you were not able to answer the first time. You may work anywhere in each section.
4. If you cannot answer a question, code it and go on to the next. Do not spend a lot of time on one question unless you have already finished the rest of that section. Go through each section and do the easiest questions first, then go back to the difficult ones.
5. Remember to use the coding system.

CLEP REVIEW—MATHEMATICS

The sample test questions are very similar to the types of questions you will see on the actual examination, although the test will be administered on-line in most test locations. Through practice on the sample test you should be able to recognize the types of problems which are easier for you and those which are more difficult. This will result in the highest degree of efficiency for your time during the examination.

The problems on the examination are not in order of difficulty. It is especially important that you make an attempt to get through every problem. Using the coding system described earlier will help you develop a strategy for working through the test. Also, there is often a recall factor in these tests. Something you read in a later problem may give you the clue to an earlier problem.

Outline of CLEP Examination

The examination consists of approximately 60 questions to be answered in 90 minutes. This gives you an average of one and a half minutes to answer each question. However, some of the questions are more difficult and require much more time, while others require much less time. The examination consists of the following topics and the approximate number of problems:

set theory	6 questions
logic	6 questions
number system	12 questions
functions and graphs	12 questions
probability and statistics	15 questions
algebra & geometry	9 questions
TOTAL	60 questions

These are topics usually covered in college mathematics courses for non-mathematics majors. The set theory questions will test your knowledge of the relationships between sets of objects, using the symbols associated with sets. In logic you will be using some new symbols to replace words which express mathematical concepts. Problems about the number system test your understanding of the different kinds of numbers in our system. An important concept, which leads to more difficult areas, is functions. Probability and statistics will utilize many familiar words in the mathematical meaning. Other topics include logarithms, complex numbers, and other number bases. These are some of the more difficult topics, and you should not spend too much time on these until you have a grasp of the other areas.

PROPERTIES OF NUMBERS

Commutative

It doesn't matter in what order we add or multiply two numbers, the sum or product will be the same.

- $3 + 5 = 5 + 3$ In both cases, the sum is 8. This is called the *Commutative Property of Addition.*

  $3 \times 5 = 5 \times 3$ In both cases, the product is 15. This is called the *Commutative Property of Multiplication.*

Associative

If we are adding or multiplying two or more numbers together, we can combine any two at a time, and the final result will be the same.

- $(3 + 5) + 8 = 3 + (5 + 8)$
  $8 + 8 = 3 + 13$
  $16 = 16$

  In both cases, the sum is 16. This is called the *Associative Property of Addition.*

- $(3 \text{ x } 5) \text{ x } 8 = 3 \text{ x } (5 \text{ x } 8)$
  $15 \text{ x } 8 = 3 \text{ x } 40$
  $120 = 120$

In both cases, the product is 120. This is called the *Associative Property of Multiplication.*

Distributive

The *Distributive Property* combines both addition and multiplication while also introducing a new way to express multiplication. When you write a number directly next to a set of parentheses, you multiply everything inside the parentheses by that number.

- $2(3 + 2 + 5) = 2(10)$
  $= 20$

We can add the numbers in the parentheses together and then multiply the sum by 2.

Or distribute the 2 to each component inside the parentheses:

- $2(3) + 2(2) + 2(5) = 6 + 4 + 10$
  $= 20$

Order of Operations

Unless numbers are grouped together with brackets, you always perform multiplications and divisions before additions and subtractions.

♦ Simplify the expression: $16 \div 4 + 4 \times 2$

(A) 12
(B) 16
(C) 4
(D) 1

Remember, we must perform division and multiplication first.

$$16 \div 4 \;+\; 4 \times 2 \;=$$
$$4 \;+\; 8 \;=\; 12$$

The correct answer is (A).

Number Relationships

Integers are whole numbers, both positive and negative: -6, -3, -1, 1, 7, 23, 48 and 526 are all examples of integers. Integers that are divisible by 2 are even integers. Those not divisible by 2 are odd, except for 0. Groups of integers can be organized into sets and are represented with brackets around the list of members included in the set.

The set {7, 8, 9, 10} represents a set of 4 consecutive integers.

The set {12, 14, 16, 18, 20} represents a set of 5 consecutive <u>even</u> integers.

There are some basic rules for computations with odd and even integers. Questions about odd and even numbers are often found on the CLEP exam. Below is a table that summarizes the relationship between even and odd numbers, followed by a couple examples of odd and even integer questions typically found on the CLEP exam:

addition	odd	even
odd	= even	= odd
even	= odd	= even

multiplication	odd	even
odd	= odd	= even
even	= even	= even

♦ If X represents an even number and Y represents an odd number, which of the following are not odd?

I. X + Y + 1
II. 2Y + 3
III. XY + Y

(A) I & II
(B) II & III
(C) I only
(D) II only

Referring to the table above, the only way to get an odd number in addition is if one number is even and the other number is odd. In multiplication, the only way to get an odd number is if both numbers are odd.

In examining the three possibilities:

I is even because you have the sum of an odd and an even which is odd. Adding one more makes it even. This means that I must be in the answer; this eliminates (B) and (D) as possibilities.

II is odd because the product of 2 and Y must be even, and the sum of an even and an odd is odd. This means that II cannot be in the answer; this eliminates (A).

III is odd because the product of an odd and an even is even; the sum of an even and an odd is odd.

During the actual test, you should not have taken the time to evaluate option III, because the answer had to be (C) by process of elimination. If you wanted to double-check and verify that III was not even, you should have answered the question with (C) and marked it for later review. If you had time left after answering all the other questions on the test, then you could review option III.

♦ Which of the following must be odd?

I. The product of two odd numbers
II. The sum of an odd and an even number
III. The product of two even numbers

(A) I only
(B) II only
(C) I and II only
(D) I and III only

In examining the three possibilities:

I is odd because the product of two odd numbers is odd. We can eliminate choice (B) because it does not include option I.

II is odd because the sum of an odd and even number is odd. We can eliminate choices (A) and (D) because they do not include option II.

By process of elimination, (C) is the correct answer.

As above, do not take the time to do further analysis during your first pass through the test. Mark the answer for further review if you have time left at the end of the test to check option III.

III is even because the product of two even numbers is even, so III is not part of our solution.

There may be questions that test your knowledge and understanding of odd and even numbers and their relationship with exponents. If an exponent is involved in an expression, remember that an odd number with any exponent produces an odd number and an even number with any exponent produces an even number. So, if a is odd then a^3, a^8 and a^{10} are also odd. If a is even then a^2, a^7 and a^{12} are also even.

♦ If $a = b^2$, what happens to a when b is tripled?

In this kind of problem you can see the result by substituting 3b for b. Remember to square 3b: $a = 3(b^2)$

$A = 9b^2$

Therefore, a is nine times larger when b is tripled or, as you might see, $3^2 = 9$.

A problem to test your understanding of the effect that exponents have on fractions might be as follows:

♦ If $P < 0$, what is true of P^3?

(A) $P^3 > P$
(B) $P^3 < P$
(C) $P^3 = P$
(D) $P^3 = P^2 + P$
(E) cannot be determined from given information

Notice that it does not say that P is a whole number, only that P is negative. If P were a whole number, then raising a negative whole number to the third power would make a smaller or equal negative number.

Substituting -2 for P would result in: $(-2)^3 = -8$, a smaller negative number.

If P were a negative fraction greater than -1, then raising it to the third power would make a larger negative number. Substituting $-\frac{1}{2}$ for P would result in $\left(-\frac{1}{2}\right)^3 = -\frac{1}{8}$ which is greater (closer to zero) than $-\frac{1}{2}$. If P were equal to -1, then P^3 would also be -1.

These examples have eliminated answer choices (A), (B), and (C).

Choice D can also be eliminated since it violates the laws of exponents.

None of these choices is correct for all cases, and the correct answer is E.

Some problems might ask you to deal with an algebraic expression and decide if it is odd or even.

♦ If a and b are positive odd numbers, is the expression $a^2 + 3ab + b^2$ odd or even?

Since a is odd, a^2 is odd. Since b is odd, b^2 is odd.

Since a and b are odd, 3ab would be 3 times an odd, or odd. Therefore, we would have the sum of 3 odd numbers which would be odd.

Signed Numbers

Integers can be both positive and negative. Integers include all the numbers that we know as whole numbers, but in addition, they also include the opposites of these numbers - the negative numbers.

```
____|____|____|____|____|____|____|____|____|____|____|____|____|
   -6   -5   -4   -3   -2   -1    0    1    2    3    4    5    6
```

Numbers get larger as you move to the right on the number line, smaller as you move to the left.

The rule for multiplying and dividing two signed numbers is:

- ♦ when the signs are the same, the answer is positive
- ♦ when the signs are different, the answer is negative

The rules for adding and subtracting signed numbers are:

- ♦ when the signs are the same, add the numbers together and take the common sign.
- ♦ when the signs are different, take the difference of the 2 numbers and the sign of the larger.

Absolute Value

Absolute value of a number is the distance that number is away from zero on the number line. Direction on the number line does not matter in determining absolute value, so even a negative number is a positive number of units away from zero. This value is represented by the symbol "| |".

♦ Which is larger, | - 24 | , (8) (3), or (- 6) (- 4)?

(A) | - 24 |
(B) (8) (3)
(C) (- 6) (- 4)
(D) they are all equal

In all cases, the value is 24. The answer is (D)

Prime Numbers

A number is a **prime number** if it can only be divided evenly by itself and the number 1. The different combinations of numbers that equal a number when multiplied are called factors. So a number is prime if it has only two factors; 1 and itself. All even numbers can be divided by 2, so there are no even prime numbers except for the number 2 itself. An example of an odd prime number is 7. There are no factors other than 7 and 1 that can result in 7.

All numbers can be reduced into their smallest possible factors. These are called the prime factors of that number. For example, the prime factors of 20 are 2, 2, and 5. They are prime numbers that will divide evenly into 20. Twenty is not considered a prime number because there are several combination of other numbers that can result in 20; 4 and 5, 2 and 10.

♦ After the numbers 11 and 13, what is the next set of prime numbers that differ by 2?

(A) 19 and 21
(B) 51 and 53
(C) 27 and 29
(D) 17 and 19

Since 21, 27, and 51 are not prime, choices (A), (B), and (C) can be eliminated immediately. The correct answer is (D).

Mean, Median and Mode

There are three different terms that are used when dealing with a group of numbers:

- mean: the mean is the numerical average of a group of numbers. To find the mean of a group of numbers, add the numbers together and divide by the total number of numbers in the group
- median: the median is the middle point of a set. To find the median of a group of numbers, arrange the numbers in ascending order and find the "middle". If there is no exact middle, then add the two middle numbers together and divide that sum by 2 to determine the median.
- mode: the mode of a group of numbers is the number that appears most often. A group of numbers can have more than one mode, and can have no mode (all members of the group are the same).

♦ A 10 question quiz resulted in the following scores. Find the mean, median and mode of the test scores from the following set of numbers:

{3, 4, 5, 5, 6, 7, 7, 7, 10}

mean: 3 + 4 + 5+5+6+7+7+7+ 10 = 54; 54 ÷ 9 = 6

median: (3 4 5 5) 6 (7 7 7 10); 6 is the middle number in the list, the median. There are exactly 4 numbers below it and 4 numbers above it.

mode: looking at the list of numbers, it is clear that 7 is the number that appears the most often

There may be questions like the one below; be sure to read the whole question to determine what they are asking for. In this example, you need to determine the mean and the mode, and subtract to get the correct answer.

♦ The difference between the mean and the mode of the numbers: 6, 10, 11, 16, 11, 19, and 11 is:

(A) 11
(B) 12
(C) 1
(D) 0

The mean of these numbers is equal to the sum, 84, divided by 7, which equals 12. The mode is the most common value, which is 11. The difference between these two numbers is 1. The correct answer is (C). (D) is the difference between the median and the mode.

There may also be times when you are asked to find the average of algebraic expressions. Just follow the same procedure you would if you were using numbers.

♦ Find the average of (x + 3), x - 6 and 4x - 3:

Add the group of numbers: (x + 3) + (x - 6) + (4x - 3) = 6x - 6

Then divide by 3: $\frac{6x-6}{3} = 2x-2$

In some average problems, the numbers may represent different quantities or averages themselves. In this case we have a weighted average. There will be a couple of preliminary steps in these types of problems.

♦ If 8 shirts of one kind cost $10.00 each and 4 of another kind cost $16.00 each, what is the average cost of all the shirts?

Here we must multiply to find the total sum first:

$8 \times 10 = \$80.00$

$4 \times 16 = \$64.00$

Total sum is $80 + $64 = $144.00; 12 shirts cost $144.00. Therefore, the average for all 12 is $12.00.

You can see that if we simply added the two pieces together to get the average, we would end up with an incorrect answer:

$$\frac{10+16}{2} = \frac{26}{2} = \$13.00$$

Some problems might give you an average and ask you to find the missing score.

♦ If John's scores are 70, 84, 74 and 76, what must he get on his next test to average 80 for the five tests?

Total sum required = average desired (80) × number of tests (5).

Total points already accumulated = 70 + 84 + 76 + 74 = 304.

(Total required) minus (total accumulated): 400 - 304 = 96 must be the score on next test.

PRACTICE—PROPERTIES OF NUMBERS

1. If a is an integer, which of the following is (are) also?

   A. $\frac{2a+2}{2}$

   B. $\frac{a+2}{2}$

   C. $\frac{2a+1}{2}$

2. If a is even, is xyab even or odd where x, y, and b are integers?

3. What happens to c when d is doubled if $c = 3d^2$?

4. A perfect number is any number equal to the SUM of its factors, including 1 but excluding the number itself, what is the first perfect number?

5. If x is odd and y is even, is (xy) + (x + y) odd or even?

6. If P is a negative fraction, which is larger P^2 or P^3?

7. If a is odd and b is even, is $a^3 \times b^3$ odd or even?

8. Is the square of an even number divided by another even number always odd, always even or may it be either?

9. 10 shirts cost $12 each, 8 shirts of another kind cost $15 each. What is the average of all 18 shirts?

ANSWERS – PROPERTIES OF NUMBERS

1. A. $\frac{2a+2}{2} = \frac{2(a+1)}{2} = a+1$ Therefore, A is an integer.

   B. $\frac{a+2}{2} = \frac{a}{2} + \frac{2}{2} = \frac{a}{2} + 1$ Therefore, B is an integer only when a is even.

   C. $\frac{2a+1}{2} = \frac{2a}{2} + \frac{1}{2} = a + \frac{1}{2}$ Therefore, C is not an integer.

2. a is even; therefore, any number of factors will produce an even integer (assuming x, y and b are integers).

3. $c = 3d^2 \quad c = 3(2d)^2$

   $c = 3 \times 4d^2$

   $c = 12d^2$

   Therefore, c is 4 times (or 2^2) larger.

4. The first perfect number is 6, since 6 = 3 + 2 + 1. The next perfect number would be 28 because 28 = 14 + 7 + 4 + 2 + 1.

5. Odd, because xy is even, xy + y is odd, and even + odd = odd.

6. P is negative. Therefore, P^3 is negative. P^2 is positive, so $P^2 > P^3$.

7. Even, because a^3 is odd, b^3 is even, and odd × even = even.

8. Since an odd number multiplied by an even produces an even number and so does an even multiplied by an even, it is impossible to determine the quotient of two even numbers.

   EXAMPLE: $\frac{6^2}{4} = \frac{36}{4} = 9(\text{odd}), \ \frac{8^2}{4} = \frac{64}{4} = 16(\text{even})$

9. 10 shirts @ \$12 = \$120

   8 shirts @ \$15 = 120

   Therefore, 18 shirts = \$240

   240 ÷ 18 = \$13.33

FRACTIONS

A **proper fraction** represents part of a whole; the numerator is smaller than its denominator, and its value is less than 1.

♦ $\frac{5}{6}, \frac{7}{9}, \frac{2}{3}, \frac{17}{18}$ are proper fractions.

An **improper fraction** has a numerator that is larger than or equal to the denominator; it has a value greater than or equal to 1.

♦ $\frac{6}{5}, \frac{9}{7}, \frac{5}{1}, \frac{4}{4}$ are improper fractions.

A **mixed number** is a whole number and a fraction. In working with mixed numbers, sometimes it is easier to change the mixed number to an improper fraction to do calculations.

♦ $3\frac{1}{4} = \frac{3 \cdot 4 + 1}{4} = \frac{13}{4}$

or: 3 can also be written as $\frac{12}{4}$, plus $\frac{1}{4}$, which equals $\frac{13}{4}$

Equivalent fractions are fractions that are equal in value. When equivalent fractions are reduced to lowest terms, they will be exactly the same.

♦ $\frac{3}{12}, \frac{6}{24}, \frac{8}{32}$ are equivalent fractions. They all reduce to $\frac{1}{4}$.

A **complex fraction** is a fraction whose numerator or denominator, or both, are fractions, or equations with fractional components. To simplify complex fractions, divide the fractions. There may be many steps in simplifying a complex fraction; don't get overwhelmed. Take it one step at a time; these are basic math skills.

♦ $$\frac{2-\frac{3}{5}}{1+\frac{1}{3+\frac{1}{4}}} = \frac{\frac{10}{5}-\frac{3}{5}}{1+\frac{1}{3\frac{1}{4}}} = \frac{\frac{7}{5}}{1+\frac{1}{\frac{13}{4}}} = \frac{\frac{7}{5}}{1+\frac{4}{13}} = \frac{\frac{7}{5}}{\frac{17}{13}} = \frac{7}{5} \div \frac{17}{13} = \frac{7}{5} \times \frac{13}{17} = \frac{91}{85}$$

Operations with Fractions

- To add and subtract fractions, the fractions must have the same denominator. Once the fractions have the same denominator, then just add or subtract the numerators, and reduce or simplify the resulting fraction, if possible.
- When adding or subtracting mixed numbers, the numbers can be changed to improper fractions and changed to common denominators; or the problem can be worked in two pieces. You can do the operation (addition or subtraction) on the whole numbers first, and on the fractional part second. Remember to add them back together for the final solution. If you choose the second method, be sure to check that you don't need to "borrow" one for the fractional subtraction before working with the two parts of the numbers.

  For example:

  $3\frac{3}{8}-1\frac{5}{8}$; you cannot subtract $\frac{5}{8}$ from $\frac{3}{8}$, so the first value must be changed to $2\frac{11}{8}$ before the calculation can take place.
- To multiply fractions, multiply the numerators together, multiply the denominators together, and then reduce, if possible. Sometimes it is possible to reduce before multiplying. This is called canceling. It is simply the process of pre-dividing any numerator in the equation with any denominator in the equation.
- To divide fractions, invert the divisor (second fraction), then proceed as in multiplication.
- With mixed numbers, for both multiplication and division, change the mixed number to an improper fraction and proceed with normal multiplication and division rules for fractions.

With any type of question, be sure to check the answers before doing any modification of your answer (i.e. improper fraction to mixed number). The answer may only ask for the improper fraction, and you don't want to waste time doing a step that is not necessary.

Below are some examples with strategies on answering fraction questions more quickly.

◆ Which is the largest fraction; $\frac{7}{12}$, $\frac{7}{15}$, or $\frac{5}{12}$?

(A) $\frac{7}{12}$

(B) $\frac{7}{15}$

(C) $\frac{5}{12}$

(D) they are all equal

Knowing a couple of simple facts will help you answer this question without spending all the time calculating common denominators for each fraction.

If the numerator remains constant, the value of a fraction becomes smaller as you increase the denominator. So, $\frac{7}{12}$ is larger than $\frac{7}{15}$. If the denominator remains constant, then the value of a fraction becomes smaller as you decrease the numerator. So $\frac{7}{12}$ is larger than $\frac{5}{12}$. The correct answer is (A).

◆ Arrange the numbers, $3\frac{2}{3}$, $\frac{17}{5}$, and $3\frac{4}{5}$, in ascending order.

(A) $3\frac{2}{3}$, $\frac{17}{5}$, $3\frac{4}{5}$

(B) $3\frac{2}{3}$, $3\frac{4}{5}$, $\frac{17}{5}$

(C) $\frac{17}{5}$, $3\frac{2}{3}$, $3\frac{4}{5}$

(D) $3\frac{4}{5}$, $\frac{17}{5}$, $3\frac{2}{3}$

For comparison purposes, it's best to change $\frac{17}{5}$ to a mixed number ($3\frac{2}{5}$). We know that $3\frac{2}{3}$ is less than $3\frac{4}{5}$, because as we add 1 to both the numerator and denominator, we increase the value of the number. So, we can eliminate

(D). $3\frac{2}{5}$ is less than $3\frac{2}{3}$ because the numerators are the same and 5 is greater than 3. So, we can eliminate (A) and (B). The correct answer is (C)

- Find the sum of $3\frac{2}{3}$ and $2\frac{1}{8}$.

(A) $5\frac{3}{11}$

(B) $6\frac{1}{24}$

(C) $\frac{77}{24}$

(D) $5\frac{19}{24}$

Again, with estimation, we have a number a little greater than $3\frac{1}{2}$ that we're adding to a number a little larger than 2. We expect a number between $5\frac{1}{2}$ and 6, maybe a little larger. This easily eliminates (A) and (C) as potential answers. Also, $\frac{1}{8}$ is less than $\frac{1}{3}$, so we know that the sum of $\frac{1}{8}$ and $\frac{2}{3}$ is less than 1. So, $3\frac{2}{3}$ plus $2\frac{1}{8}$ is less than 6. (D) must be the right answer.

PRACTICE PROBLEMS - FRACTIONS

1. $\frac{3}{4}+2\frac{5}{7}=$

2. $\frac{8}{9}-\frac{2}{5}=$

3. $3\frac{2}{3}\times 4\frac{1}{2}=$

4. $\frac{3}{4}\div 5\frac{1}{2}=$

5. $\dfrac{\frac{5}{6}}{\frac{2}{3}}=$

6. Reduce $\frac{39}{57}$

7. $17\frac{5}{8}\times 128=$

8. $\frac{7}{8}\times\frac{2}{3}\div\frac{4}{7}=$

9. $12\frac{1}{3}-6\frac{3}{4}=$

10. $\frac{5}{6}+\frac{2}{5}-\frac{1}{10}=$

11. $3\frac{2}{3}\div 1\frac{1}{6}=$

12. $\left(\frac{5}{6}+\frac{1}{3}\right)\times\left(\frac{2}{3}+\frac{3}{4}\right)=$

13. $\frac{5}{8}\times\frac{2}{3}\div\frac{5}{6}=$

14. $7\frac{1}{2}\left(3\frac{1}{4}+5\frac{1}{5}\right)=$

15. $\left(\frac{2}{3}\div\frac{5}{8}\right)\div 1\frac{1}{2}=$

16. $2+\dfrac{1}{2+\frac{1}{2}}=$

ANSWERS – FRACTION PROBLEMS

1. $\frac{3}{4}+2\frac{5}{7}=\frac{3}{4}+\frac{19}{7}=\frac{3\cdot 7+19\cdot 4}{4\cdot 7}=\frac{21+76}{28}=\frac{97}{28}=3\frac{13}{28}$

2. $\frac{8}{9}-\frac{2}{5}=\frac{8\cdot 5-2\cdot 9}{9\cdot 5}=\frac{40-18}{45}=\frac{22}{45}$

3. $3\frac{2}{3}\times 4\frac{1}{2}=\frac{11}{\not{3}_1}\times\frac{\not{9}^3}{2}=\frac{33}{2}=16\frac{1}{2}$

4. $\frac{3}{4}\div 5\frac{1}{2}=\frac{3}{4}\div\frac{11}{2}=\frac{3}{\not{4}_2}\times\frac{\not{2}^1}{11}=\frac{3}{22}$

5. $\dfrac{\frac{5}{6}}{\frac{2}{3}} = \dfrac{5}{6} \div \dfrac{2}{3} = \dfrac{5}{{}_2\cancel{6}} \times \dfrac{\cancel{3}^{1}}{2} = \dfrac{5}{4} = 1\dfrac{1}{4}$

6. $\dfrac{39}{57} = \dfrac{39 \div 3}{57 \div 3} = \dfrac{13}{19}$

7. $17\dfrac{5}{8} \times 128 = 17(128) + \dfrac{5}{{}_1\cancel{8}}(\cancel{128}^{16}) = 2{,}176 + 80 = 2{,}256$

8. $\dfrac{7}{8} \times \dfrac{2}{3} \div \dfrac{4}{7} = \dfrac{7}{{}_4\cancel{8}} \times \dfrac{\cancel{2}^{1}}{3} \times \dfrac{7}{4} = \dfrac{7 \cdot 7}{4 \cdot 3 \cdot 4} = \dfrac{49}{48} = 1\dfrac{1}{48}$

9. $12\dfrac{1}{3} - 6\dfrac{3}{4} = \dfrac{37}{3} - \dfrac{27}{4} = \dfrac{37 \cdot 4 - 27 \cdot 3}{3 \cdot 4} = \dfrac{148 - 81}{12} = \dfrac{67}{12} = 5\dfrac{7}{12}$

10. $\dfrac{5}{6} + \dfrac{2}{5} - \dfrac{1}{10} = \dfrac{25}{30} + \dfrac{12}{30} - \dfrac{3}{30} = \dfrac{25 + 12 - 3}{30} = \dfrac{34}{30} = 1\dfrac{4}{30} = 1\dfrac{2}{15}$

11. $3\dfrac{2}{3} \div 1\dfrac{1}{6} = \dfrac{11}{3} \div \dfrac{7}{6} = \dfrac{11}{{}_1\cancel{3}} \times \dfrac{\cancel{6}^{2}}{7} = \dfrac{22}{7} = 3\dfrac{1}{7}$

12. $\left(\dfrac{5}{6} + \dfrac{1}{3}\right) \times \left(\dfrac{2}{3} + \dfrac{3}{4}\right) = \left(\dfrac{5}{6} + \dfrac{2}{6}\right) \times \left(\dfrac{8}{12} + \dfrac{9}{12}\right) = \dfrac{7}{6} \times \dfrac{17}{12} = \dfrac{119}{72} = 1\dfrac{47}{72}$

13. $\dfrac{5}{8} \times \dfrac{2}{3} \div \dfrac{5}{6} = \dfrac{5}{8} \times \dfrac{2}{3} \times \dfrac{6}{5} = \dfrac{\cancel{5}^{1}}{{}_4\cancel{8}} \times \dfrac{\cancel{2}^{1}}{{}_1\cancel{3}} \times \dfrac{\cancel{6}^{2}}{\cancel{5}_1} = \dfrac{2}{4} = \dfrac{1}{2}$

14. $7\dfrac{1}{2}\left(3\dfrac{1}{4} + 5\dfrac{1}{5}\right) = 7\dfrac{1}{2}\left(\dfrac{13}{4} + \dfrac{26}{5}\right) = 7\dfrac{1}{2}\left(\dfrac{65}{20} + \dfrac{104}{20}\right) = 7\dfrac{1}{2}\left(\dfrac{169}{20}\right) = \left(\dfrac{15}{2}\right)\left(\dfrac{169}{20}\right) =$

$\dfrac{15 \times 169}{40} = \dfrac{2535}{40} = 63\dfrac{15}{40} = 63\dfrac{3}{8}$

15. $\left(\dfrac{2}{3} \div \dfrac{5}{8}\right) \div 1\dfrac{1}{2} = \dfrac{2}{3} \times \dfrac{8}{5} \times \dfrac{2}{3} = \dfrac{32}{45}$

16. $2 + \dfrac{1}{2 + \frac{1}{2}} = 2 + \dfrac{1}{2\frac{1}{2}} = 2 + \dfrac{1}{\frac{5}{2}} = 2 + \dfrac{2}{5} = 2\dfrac{2}{5}$

DECIMALS

Decimals are a form of fraction where the denominator is understood to be a power of 10. The first number to the right of the decimal indicates the 10ths place or 10^{-1}, the second the 100ths place or 10^{-2}, the third the 1,000ths place or 10^{-3}, etc.

- .37 is the equivalent of $\frac{37}{100}$

  .024 is the equivalent of $\frac{24}{1000}$

Note that adding zeros to the right of both the numerator and denominator does not change the value: $.8 = \frac{8}{10} = \frac{80}{100} = \frac{800}{1000}$, etc.

A whole number with a decimal component can be written as a mixed number.

- $38.35 = 38\frac{35}{100}$

Operations with Decimals

- To add and subtract decimals, line up the decimal points and bring the decimal point straight down. Then add or subtract the numbers as if they were whole numbers. If there is a whole number without a decimal, just add a decimal point with zeroes after it.
- You do not need to line up the decimal points for multiplication. Multiply as if the numbers were whole numbers without decimals. Count the total number of decimal places in the factors, and add a decimal point in the answer after that many positions, counting from right to left.

$$\begin{array}{r} 6.25 \\ \times 3.2 \\ \hline 1250 \\ 18750 \\ \hline 20.000 \end{array}$$

> Counting the number of decimal places in the factors gives a total of 3 places. Take the answer, 20000, and mark off three decimal places, beginning at the end of the number. The final answer would then be 20.000, or just 20.

- Dividing decimals involves changing the divisor to a whole number. This is done by multiplying both the divisor and the dividend by the power of 10 that will clear the decimal place from the divisor. If there are still decimal places left in the dividend, then move the decimal point straight up on the quotient answer line, and divide the numbers normally. If there are not any decimal places left in the dividend after the multiplication, you may need to add a decimal point and a few zeroes to complete the division problem.

$$62.50 \div 2.5 = 25\overline{)625.}$$

25. 25)625. 50 125 125 0	The divisor, 2.5, must be multiplied by 10 to clear the decimal place. So the dividend must also be multiplied by 10, giving 625. Continue with normal division.

♦ To change a fraction to a decimal, divide the denominator of the fraction into the numerator. This is a case when the decimal point and additional zeroes must be added to the dividend.

$$\frac{3}{8} = 3 \div 8 = 8\overline{)3.000}$$

0.375 8)3.000 2.4 .600 .560 .040 .040 .000	In order to get the decimal answer, as a rule of thumb, add a decimal point and 3 zeroes after the numerator. You may get a repeating decimal – one that does not end. In this case, it may be best to use the fraction form for computations.

When adding multiple types of numbers together, decide on one format, and convert all numbers to that format to complete your calculations. Be sure to check the answer choices to help decide which format to use. If the answers are in fraction form, then be sure to use fractions in your calculations to save time.

PRACTICE PROBLEMS - DECIMALS

1. $.08 \times 7.2 =$

2. $25 + 3.24 + .04 =$

3. Change to fractions:

   A. .12

   B. 1.2

   C. .58

   D. .002

4. Change to decimals:

   A. $\frac{5}{8}$

   B. $\frac{13}{15}$

   C. $\frac{8}{7}$

   D. $3\frac{1}{4}$

5. Write 12.42 as a mixed number:

6. $.25\overline{)6.25}$

7. $\frac{4}{5} + .72 =$

8. Write $\frac{5.6}{100}$ as a decimal:

9. $(2.3 \times 4) + (.025 \times 6.2) =$

10. $\frac{1}{5} \times .035 =$

11. $(.3)\left(\frac{1}{2}\right) + (.05)\left(\frac{4}{5}\right) =$

12. $.27\overline{)1.215}$

13. Add $\frac{2}{5}$, .05, and 1.2. Express the answer in decimal form.

14. Add $\frac{2}{5}$, .07, and $\frac{5}{8}$. Express the answer in decimal form.

ANSWERS - DECIMALS

1. $.08 \times 7.2 = .576$

2. $25 + 3.24 + .04 = $

$$\begin{array}{r} 25.00 \\ 3.24 \\ +\ .04 \\ \hline 28.28 \end{array}$$

3. A. $.12 = \frac{12}{100} = \frac{3}{25}$

   B. $1.2 = 1\frac{2}{10} = 1\frac{1}{5} = \frac{6}{5}$

   C. $.58 = \frac{58}{100} = \frac{29}{50}$

   D. $.002 = \frac{2}{1000} = \frac{1}{500}$

4. A. $\frac{5}{8} = 8\overline{)5.000}$ = .625

   B. $\frac{13}{15} = 15\overline{)13.00}$ = .866...

   C. $\frac{8}{7} = 7\overline{)8.00}$ = $1.142\frac{6}{7}$

   D. $3\frac{1}{4} = \frac{13}{4} = 4\overline{)13.00}$ = 3.25

5. $12.42 = 12\frac{42}{100} = 12\frac{21}{50}$

6. $25\overline{)625}$ = 25.

$$\begin{array}{r} 50 \\ \hline 125 \\ 125 \\ \hline 0 \end{array}$$

7. $\frac{4}{5} + .72,\ 5\overline{)4.00}$ = .80, $.8 + .72 = 1.52$

8. $\frac{5.6}{100} = \frac{56}{1000} = .056$

9. $(2.3 \times 4) + (.025 \times 6.2) = 9.2 + .1550 = 9.3550$

10. $\frac{1}{5} \times .035 = \frac{.035}{5} = 5\overline{).035}$ = .007

11. $(.3)\left(\frac{1}{2}\right)+(.05)\left(\frac{4}{5}\right)=\left(\frac{3}{10}\right)\left(\frac{1}{2}\right)+\left(\frac{\overset{1}{\not 5}}{\underset{25}{\not{100}}}\right)\left(\frac{\overset{1}{\not 4}}{\underset{1}{\not 5}}\right)=\frac{3}{20}+\frac{1}{25}=\frac{15}{100}+\frac{4}{100}$

$=\frac{19}{100}=.19$ OR $(.3)(.5) + (.05)(.8) = .15 + .04 = .19$

12. $.27\overline{)1.215} = 27\overline{)121.5}$

$$\begin{array}{r} 4.5 \\ 27\overline{)121.5} \\ \underline{108} \\ 13\,5 \\ \underline{13\,5} \\ 0 \end{array}$$

13. $\frac{2}{5}+.05+1.2=.4+.05+1.2=1.65$

14. $\frac{2}{5}=.4$ and $\frac{5}{8}=.625$

Together:

$$\begin{array}{r} .400 \\ .070 \\ \underline{.625} \\ 1.095 \end{array}$$

PERCENTS

Decimals & Percents

Percent is another way you can refer to a part of 100. 1% is 1 part of 100. 100% is 100 parts of 100, or the entire quantity. The decimal .25 is 25 parts of 100 which is 25%

Changing from decimal form to percent form, shift the decimal point two places to the right (multiply by 100) and add the percent symbol. Changing from percent to decimal form, shift the decimal point two places to the left (divide by 100) and drop the percent symbol.

- ♦ To change .093 to a percent, shift the decimal 2 places to the right and add a percent sign; 9.3%.
- ♦ To change 6.4% to a decimal, shift the decimal 2 places to the left and remove the percent sign; adding a zero in front of the 6 yields .064.

Fractions & Percents

Changing from fraction form to percent form, you have the added step of changing the fraction to a decimal. Then you proceed as before.

Changing from percent to fraction form is a matter of dividing by 100 and dropping the percent symbol. If there is a fraction in the percent, you must change the number to an improper fraction or change the fractional portion to a decimal before dividing by 100.

♦ Change $3\frac{1}{4}\%$ to a fraction

You must first change $3\frac{1}{4}\%$ to $\frac{13}{4}\%$ or 3.25%

Dividing by 100, we would multiply the fraction $\frac{13}{4}$ by $\frac{1}{100}$, or shift the decimal 2 places to the left:

$\frac{13}{4}\% = \frac{13}{400}$ $\quad\quad$ $3.25\% = .0325 = \frac{325}{10000} = \frac{13}{400}$

Percent Word Problems

There are 3 elements to a percent problem:

- ♦ the whole; the **base**
- ♦ the part of the whole that you are interested in; the **percentage**
- ♦ the percentage of the whole that this part represents; the **rate**

The percent of any two numbers can be obtained by dividing the part into the whole. So if you had 4 pieces of pie, and 3 are eaten, you can calculate the percentage of pie eaten by diving the part that was eaten (3 pieces) into the whole pie (4 pieces): 3 divided by 4: .75 = 75%.

Most percent problems can be summarized into a statement using the three pieces listed above. These statements take the following form:

- ♦ What percent of A is B?
- ♦ A is what percent of B?
- ♦ What number is some given percent of A?

Translating these sentences to equations is the next step; "of" translates to multiply by, and "is" translates to equals.

♦ What percent of 200 is 50?

200 x ?% = 50

You would solve this like any other equation; divide both sides by 200.

You would then have ?% = .25.

Changing that decimal to a percent would leave you with the answer of 25%.

♦ What is 30% of $12.50?

? = 30% x 12.50

.3 x 12.5 = 3.75.

A quick and easy way to do this problem is to take 10% of 12.50 by shifting the decimal 1 place to the left -- 10% = $1.25. 30% is 3 times larger or $3.75. This method can be used for any multiple of 10%.

The following is an example of the third type of percentage statement; when the part is given and the question is asking you to find the whole.

◆ 15% of what number is 5.40?

The part is 5.40, the percent is 15, we want to find the whole.

Divide the part by the percent:

$$.15\overline{)5.40} = 15.\overline{)540.0}\quad (36.0)$$

You should be familiar with some basic formulas involving percent problems.

In interest, the rate is stated as a percent per year. The formula to find the amount of interest is principal times rate times time:

Interest formula: $I = P \times R \times T$

◆ To find the interest on $1,200 at 6% per year for 6 months:

$$I = 1{,}200 \times .06 \times \frac{1}{2} = 72 \times \frac{1}{2} = \$36.00$$

NOTE: If a problem expresses the rate in terms of months, then the time must be calculated in terms of months and vice versa.

Be sure you read carefully to see if you are asked to find the interest or the total at the end of the period. In the above problem, the interest would be $36, but the total in the bank would be $1,200 + $36 = $1,236. Both answers would be present in the multiple choice options, so be sure to know what they are asking.

◆ Find the amount of principal needed to earn $300 a year interest at 8% for a year.

$$\begin{aligned} I &= P \times R \times T \\ 300 &= ? \times 8\% \times 1 \\ 300 &= ? \times 8\% \\ 300 \div 8\% &= ? \end{aligned}$$

$$300 \div \frac{8}{100} = \frac{300}{1} \div \frac{2}{25} = \frac{\overset{150}{\cancel{300}}}{1} \times \frac{25}{\underset{1}{\cancel{2}}} = 150 \times 25 = \$3{,}750$$

Discount problems are based on the statement:

$$\text{Amount of Discount} = \text{Rate} \times \text{Original Price}$$

♦ Find the rate of discount on an $80 coat, which is discounted to $60.

Amount of Discount = 80 - 60 = $20

Amount of Discount = Rate × Original Price

$\$20 = ? \times \$80$

$\frac{20}{80} = ?$

$\frac{1}{4} = 25\%$

Remember, if a discount rate is given, the amount paid represents 100% minus the discount rate. For example, on a discount of 30%, the amount paid represents

100% - 30% or 70%.

♦ Find the amount paid on a suit selling for $120 if a 30% discount is given.

100% - 30% = 70%

70% × 120 = $84.00

A favorite type of percent problem is related to successive discounts. You will be tested to see if you are aware that a single discount is greater than 2 successive discounts whose sum is the same as the single discount.

♦ How much more is a single discount of 40% than 2 successive discounts of 30% and 10%?

Since a 30% discount means 70% is paid, a 10% discount means 90% is paid, the successive discounts are equivalent to a payment of 90% × 70% or 63%. Therefore, the discount is 100% - 63% = 37%. A single discount of 40% is therefore 3% greater than the successive discounts of 30% and 10%.

Another common type of percent problem involves commissions.

♦ A salesman receives a salary of \$150 per week, plus a 2% commission on all sales over \$300. Find the total salary if his sales were \$2,800.

2,800 - 300 = 2,500 is the amount on which commission can be taken.

$$\text{Commission} = 2\% \times 2500 = 50.00$$

$$\text{Salary} = 150 + 50 = \$200$$

A frequent type of percent problem involves percent increase and decrease.

♦ What is the percent increase of a town whose population went from 10,000 to 12,500?

$$\text{Percent increase} = \frac{\text{amount of increase}}{\text{original amount}}$$

$$\% = \frac{12{,}500 - 10{,}000}{10{,}000} = \frac{2{,}500}{10{,}000} = \frac{1}{4} = 25\%$$

Remember in percent increase and decrease to use the original number as your denominator, not the new amount.

♦ If the attendance at a school football game went down from 2,800 to 2,100, what is the percent decrease?

$$\frac{2{,}800 - 2{,}100}{2{,}800} = \frac{700}{2{,}800} = \frac{1}{4} = 25\%$$

Some problems may deal with a percent greater than 100%. This means that the problem is dealing with an amount greater than the original.

♦ If John's salary is 130% of Frank's and John earns $31,200 per year, what is Frank's salary?

$$
\begin{aligned}
130\% \times \text{Frank's salary} &= 31{,}200 \\
\text{Frank's salary} &= 31{,}200 \div 130\% \\
&= 31{,}200 \div 1.30 \\
&= 24{,}000
\end{aligned}
$$

PRACTICE PROBLEMS - PERCENTS

Try these problems:

1. Change to fractions:

   A. 28% B. 3% C. 5.6% D. 120%

2. Change to percents:

   A. $\frac{5}{6}$ B. .024 C. $4\frac{1}{5}$ D. .002

3. Change to fractions:

   A. $3\frac{1}{2}\%$ B. $\frac{2}{5}\%$ C. 15.05%

4. 52% of 720 =

5. 80 ÷ 30% =

6. $2\frac{1}{2}\%$ of 7,200 =

7. 80% of 90% =

8. 98% ÷ 14% =

9. A. 20% of 80 =

   B. What % of 10 is 20?

   C. 20% of what number is 14?

10. $\frac{1}{2}$ is what % of $\frac{2}{3}$?

11. Which of the values below is the largest?

    (A) $\frac{.1}{.3}$

(B) $\frac{.1}{3}$

(C) $\frac{1}{.3}$

(D) 310%

12. 42 is $37\frac{1}{2}\%$ of?

13. 235% of 780 =

14. 120% of what number is 54?

ANSWERS - PERCENTS

1. A. $28\% = \frac{28}{100} = \frac{7}{25}$

   B. $3\% = \frac{3}{100}$

   C. $5.6\% = \frac{5.6}{100} = \frac{56}{1000} = \frac{7}{125}$

   D. $120\% = \frac{120}{100} = \frac{6}{5} = 1\frac{1}{5}$

2. A. $\frac{5}{6} = 6\overline{)5.00}^{\,.83...} = 83\frac{1}{3}\%$

   B. $.024 = 2.4\%$

   C. $4\frac{1}{5} = 4.2 = 420\%$

   D. $.002 = .2\%$

3. A. $3\frac{1}{2}\% = \frac{7}{2}\% = \frac{7}{200}$

   B. $\frac{2}{5}\% = \frac{2}{500} = \frac{1}{250}$

   C. $15.05\% = \frac{15.05}{100} = \frac{1505}{10000} = \frac{301}{2000}$

4. 52% of 720 =

$$\begin{array}{r} 720 \\ \underline{\times .52} \\ 1440 \\ 3600 \\ \hline 374.40 \end{array}$$

5. $80 \div 30\% = 80 \div \frac{3}{10} = 80 \times \frac{10}{3} = \frac{800}{3} = 266\frac{2}{3}$

6. $2\frac{1}{2}\%$ of 7,200

$$\begin{array}{r} 7{,}200 \\ \underline{\times .025} \\ = \ 36000 \\ 144000 \\ \hline 180.00 \end{array}$$

7. 80% of 90% = .80 or 72%

$$\begin{array}{r} .80 \\ \underline{\times .90} \\ .7200 \end{array}$$

8. $98\% \div 14\% = 14\overline{)98.00} = 700\%$ (quotient 7.00)

$$\underline{98}$$

9. A. 20% of 80 = .20 × 80 = 16.00

B. What % of 10 is 20? $\frac{20}{10} = 2 = 200\%$

C. 20% of what number is 14?

$$\text{Rate} = \frac{\text{Percentage}}{\text{Base}} = \frac{14}{20\%} = \frac{14}{1} \div \frac{1}{5} = \frac{14}{1} \times \frac{5}{1} = 70$$

10. $\frac{1}{2}$ is what % of $\frac{2}{3}$?

$$\text{Rate} = \frac{\text{Percentage}}{\text{Base}} = \frac{1}{2} \div \frac{2}{3} = \frac{1}{2} \times \frac{3}{2} = \frac{3}{4} = 75\%$$

Proof: $75\% = \frac{3}{4} \qquad \frac{3}{\cancel{4}_2} \times \frac{\cancel{2}^1}{3} = \frac{3}{6} = \frac{1}{2}$

11. The answer is C. Convert all numbers to percent and compare.

A. $\frac{.1}{.3} = 33.3\%$

B. $\frac{.1}{3} = 3.33\%$

C. $\frac{1}{.3} = 333\%$

D. 310%

12. 42 is $37\frac{1}{2}\%$ of what number?

SOLUTION: $37\frac{1}{2}\% = \frac{3}{8}$ or $37\frac{1}{2}\% = .375$

$$\frac{3}{8} \times ? = 42$$

$$? = 42 \div \frac{3}{8}$$

$$? = 42 \times \frac{8}{3}$$

$$= 112$$

$$\begin{array}{r} 112 \\ 375\overline{)42000} \\ \underline{375} \\ 450 \\ \underline{375} \\ 750 \\ \underline{750} \\ 0 \end{array}$$

13. 235% of 780 = 2.35 × 780 = 1,833.00

14. 120% × ? = 54

? = 54 ÷ 120%

? = 54 ÷ 1.2

? = 45

FRACTION/PERCENT/DECIMAL EQUIVALENTS

FRACTION	PERCENT	DECIMAL
$\frac{1}{2}$	50%	.5
$\frac{1}{3}$	$33\frac{1}{3}\%$	.33...
$\frac{2}{3}$	$66\frac{2}{3}\%$	.66...
$\frac{1}{4}$	25%	.25
$\frac{3}{4}$	75%	.75
$\frac{1}{5}$	20%	.20
$\frac{2}{5}$	40%	.40
$\frac{3}{5}$	60%	.60
$\frac{4}{5}$	80%	.80
$\frac{1}{6}$	$16\frac{2}{3}\%$	.166...
$\frac{1}{8}$	$12\frac{1}{2}\%$	.125
$\frac{3}{8}$	$37\frac{1}{2}\%$	.375
$\frac{5}{8}$	$62\frac{1}{2}\%$	.625
$\frac{7}{8}$	$87\frac{1}{2}\%$	.875

FRACTION AND DIVISION TIPS

Fraction Tips

- Any proper fraction with 9 as a denominator repeats the numerator in its decimal form.

  $\frac{2}{9} = .22...$ $\frac{5}{9} = .55...$

- If you can remember the decimal and percent equivalent of $\frac{1}{8}, \frac{1}{6}, \frac{1}{5}$ etc. you can calculate the others by multiplying.

  If $\frac{1}{8} = 12.5\%$, then $\frac{7}{8} = 12.5\% \times 7 = 87.5\%$

Divisibility Tips

- Even numbers are divisible by 2.
- A number is divisible by 3 if the sum of the digits is divisible by 3.
  - ◊ 714 is divisible by 3 since 7 + 1 + 4 = 12; which is divisible by 3
- A number is divisible by 4 if the number formed by the last two digits is divisible by 4.
  - ◊ 748 is divisible by 4 since 48 is divisible by 4.
- A number is divisible by 5 if the last digit is either 5 or 0.
- A number is divisible by 6 if it is divisible by 2 and 3.
  - ◊ 732 is divisible by 6; it is divisible by 2 because it is even, and it is divisible by 3 since 7 + 3 + 2 = 12, which is divisible by 3.
- A number is divisible by 8 if the number formed by the last three digits is divisible by 8.
  - ◊ 1,296 is divisible by 8 since 296 is divisible by 8.
- A number is divisible by 9 if the sum of its digits is divisible by 9.
  - ◊ 7,875 is divisible by 9 since 7 + 8 + 7 + 5 = 27, which is divisible by 9.

BASIC ALGEBRA

Exponents

In the expression 3^4, 3 is called the base, and 4 is called the exponent. The exponent expresses the number of times that the base should be multiplied by itself. In this case, 3 should be multiplied by itself 4 times, which equals 81.

Any number can be considered as being raised to the power of 1. Any number raised to the "0" power is equal to 1 (except for 0).

You need to know certain rules for working with exponents:

♦ When you multiply like bases, you add the exponents together. For example:

$$2^3 \quad \times \quad 2^2 \quad = \quad 2^5$$

$$(2 \times 2 \times 2) \quad \times \quad (2 \times 2) \quad = 2 \text{ to the fifth power.}$$

This only works when the bases are the same.

$$4^2 \quad \times \quad 4 \quad \times \quad 4^2 \quad = \quad 4^5$$

♦ When you divide like bases, you subtract the second exponent from the first exponent.

$$3^7 \quad \div \quad 3^4$$

$$\frac{3 \times 3 \times 3 \times 3 \times 3 \times 3 \times 3}{3 \times 3 \times 3 \times 3} = 3^3$$

When the second exponent is larger than the first exponent, you will end up with a negative answer. This indicates that the exponent is in the denominator.

$$3^4 \quad \div \quad 3^7 = 3^{-3}$$

$$\frac{3 \times 3 \times 3 \times 3}{3 \times 3 \times 3 \times 3 \times 3 \times 3 \times 3} = \frac{1}{3^3}$$

♦ When you raise an exponent to another power, you multiply the exponents.

$$(3^3)^2 = 3^3 \times 3^3 = 3^6$$

We see that the new exponent is equal to the product of the 2 exponents in our problem.

Sometimes you may need to recognize the factors of an exponential expression and reduce a number to a like-base expression to perform the calculation.

Find the product of 2^8 and 4^3.

(A) 2^{11}

(B) 2^{14}

(C) 2^{48}

(D) 8^{11}

$2^8 \times (2^2)^3 = 2^8 \times 2^6 = 2^{14}$.

The correct answer is (B).

Radicals

The opposite of powers is roots. Just as you can have powers of 2, (squared), you can have roots of 2, which are called square roots.

The square root of a number is a number that when multiplied by itself will give you the original number. It is written like this: $\sqrt{\ }$. A number is called a perfect square if it can be produced by multiplying the same number by itself.

♦ $\sqrt{49} = 7$ because 7 x 7 is 49. 49 is considered a perfect square.

There are some basic rules that apply when working with square roots. You can multiply two square roots, and their product will be under the radical sign as well. You can then simplify the expression if you end up with a perfect square.

♦ $\sqrt{20} \times \sqrt{5} = \sqrt{100} = 10$

$\sqrt{18} \times \sqrt{2} = \sqrt{36} = 6$

♦ $\sqrt{75} = \sqrt{25 \times 3} = \sqrt{25} \times \sqrt{3} = 5\sqrt{3}$

$\sqrt{75}$ is not an integer; it is called an irrational number because there is no number that is equal to the square root of 75. However, even though 75 is not a perfect square, we can simplify $\sqrt{75}$ by factoring 75 into 25 and 3. Since 25 is a perfect square, the expression can be further simplified:

When adding and subtracting, you must have the same value under the radical to perform the operation. Be sure to check and see if an expression can be simplified to have the same value under the radical sign, allowing for further simplification.

♦ Combine: $3\sqrt{2} + 4\sqrt{2} - \sqrt{2} = (3 + 4 - 1)\sqrt{2} = 6\sqrt{2}$

Some questions will require you to use the tools of square roots to derive the correct answer. In the example below, you would need to know what the product of a x b is, then list the factors of that product, then subtract them to find the correct answer.

♦ If $\sqrt{ab} = 6$, and a and b are positive integers, which of the following could not be the value of a - b?

A. 5
B. 9
C. 16
D. 4

Since $\sqrt{ab} = 6$, then ab must equal 36. The factors of 36 are 36 × 1, 18 × 2, 12 × 3, 9 × 4, and 6 × 6. By subtracting each set of factors we see that only (D) could not be possible.

Rationalizing the Denominator

As a practice, radicals should not be left in the denominator of a fraction. The process of removing the square root from a denominator is called rationalizing the denominator. The process is simple and straightforward; multiply the whole fraction by the radical that is in the denominator over itself. The value of any fraction with the same value in the numerator and denominator is always 1, so the new fraction will be equivalent, and the square root will be removed, as you will be "squaring" the square root.

PRACTICE PROBLEMS – EXPONENTS AND RADICALS

1. $12^0 \times 10^3 =$

2. $2^{-4} \times 16 =$

3. $x^3 \div x^5 =$

4. $\dfrac{1}{a^{-2}} =$

5. Simplify: $\sqrt{12} + \sqrt{20} =$

6. $(3\sqrt{2})(2\sqrt{6}) =$

7. Rationalize: $\dfrac{1}{2\sqrt{3}}$

8. Simplify: $2\sqrt{3} + \sqrt{75} - \sqrt{27}$

9. Solve for a: $3\sqrt{a-4} = 15$

ANSWERS – EXPONENTS AND RADICALS

1. $12^0 \times 10^3 = 1 \times 1000 = 1000$

2. $2^{-4} = \frac{1}{2^4} = \frac{1}{16}, \ \frac{1}{16} \times \frac{16}{1} = 1$

3. $x^3 \div x^5 = x^{-2} = \frac{1}{x^2}$ Another look: $\frac{x^3}{x^5} = \frac{x \cdot x \cdot x}{x \cdot x \cdot x \cdot x \cdot x} = \frac{1}{x^2}$

4. $\frac{1}{a^{-2}} = \frac{1}{\frac{1}{a^2}} = \frac{1}{1} \div \frac{1}{a^2} = \frac{1}{1} \times \frac{a^2}{1} = a^2$

5. Simplify: $\sqrt{12} + \sqrt{20} = (\sqrt{4} \times \sqrt{3}) + (\sqrt{4} \times \sqrt{5})$
$= 2\sqrt{3} + 2\sqrt{5}$
$= 2(\sqrt{3} + \sqrt{5})$

6. $(3\sqrt{2})(2\sqrt{6}) = 3 \cdot 2 \cdot \sqrt{2} \cdot \sqrt{6}$
$= 6 \cdot \sqrt{2} \cdot \sqrt{2} \cdot \sqrt{3}$
$= 6 \cdot 2\sqrt{3}$
$= 12\sqrt{3}$

7. Rationalize: $\frac{1}{2\sqrt{3}} \times \frac{\sqrt{3}}{\sqrt{3}} = \frac{\sqrt{3}}{2 \cdot 3} = \frac{\sqrt{3}}{6}$

8. Simplify: $2\sqrt{3} + \sqrt{75} - \sqrt{27} = 2\sqrt{3} + \sqrt{25} \cdot \sqrt{3} - \sqrt{9} \cdot \sqrt{3}$
$= 2\sqrt{3} + 5\sqrt{3} - 3\sqrt{3}$
$= 4\sqrt{3}$

9. $3\sqrt{a-4} = 15$
$\sqrt{a-4} = 5$ dividing by 3
$a - 4 = 25$ squaring both sides
$a = 29$

Algebraic Expressions

$3x + 7$ is an algebraic expression specifically known as a binomial because it has two terms. The expression $x^2 + 2x - 15$ is generally known as a polynomial, because it has many terms. Specifically it is a trinomial because it has 3 terms, an x^2 term, an x term and a constant.

When adding polynomials together remember that you can only add "like terms". It is the variable that gives the name to the algebraic term, so add only "x"s to "x"s, "x^2"s to "x^2"s, etc.

♦ $3x^2 + 7x - 9$

$+5x^2 - 12x + 13$

$8x^2 - 5x + 4$

When you multiply 2 polynomials together, remember that each term of the first polynomial must multiply each term of the second.

♦ Multiply $(x + 5)(x - 3)$.

x must multiply x and -3; 5 must multiply x and -3, then like terms can be combined.

$$x^2 - 3x + 5x - 15 = x^2 + 2x - 15$$

♦ Multiply $(x^2 + 4x - 5)(x - 7)$

$x^3 - 7x^2 + 4x^2 - 28x - 5x + 35 =$

$x^3 - 3x^2 - 33x + 35$

Factoring

Factoring just reverses the multiplication process. To write a polynomial in factored form is to write it as a product of its factors. Let's first take a look at binomials. There are 2 types of factoring problems involving binomials.

Common Factor

In a common factor problem, we are looking for a number, a variable, or both that will divide evenly into each algebraic term of the binomial. In any factoring problem, it is the first factor you should look for.

♦ Write $6x - 9$ in factored form.

The number that divides evenly into both 6 and 9 is 3. Write 3 outside the parentheses; it's the common factor. Inside the parentheses, write the result of dividing the common factor into each of the binomial terms. 3 goes into 6x, leaving 2x. 3 goes into - 9, leaving - 3.

$$6x - 9 = 3(2x - 3)$$

♦ Write $5x^5 - 25x^3$ in factored form.

The number that divides evenly into both 5 and 25 is 5. The variable that divides evenly into both x^3 and x^5 is x^3. We write $5x^3$ outside the parentheses; it's the common factor. Inside the parentheses, we write the result of dividing the common factor into each of the binomial terms.

$$5x^5 - 25x^3 = 5x^3 (x^2 - 5)$$

Difference of 2 Perfect Squares

The second type of binomial factoring problem you might see is the difference of 2 perfect squares.

♦ Write $x^2 - 49$ in factored form.

From what we know about multiplying binomials together, we can get x by multiplying x by x. Similarly, we can get - 49 by multiplying - 7 by + 7. Because the number is the same in both factors, but the signs are different, the x-terms will cancel out, leaving us with just the difference between the square of the first term and the square of the second term.

$$x^2 - 49 = (x + 7) (x - 7)$$

When factoring the difference of two perfect squares, the solution is found by taking the square root of both terms, with one factor being positive and the other negative. When factoring, be sure to continue to simplify each factor into lowest terms.

♦ Write $x^4 - 81$ in factored form.

The first step to this problem would be to recognize the two perfect squares of x^4 and 81. Once the expression has been factored into two binomials, the next step is to factor again because there is another perfect square. The final answer will contain three factors:

step 1: $(x^2 + 9)(x^2 - 9) =$

step 2: $(x^2 + 9)(x + 3)(x - 3)$

Other questions will combine both methods of factoring:

♦ The correct factors of $3x^2 - 48$ are:

(A) $3(x + 4)(x - 4)$
(B) $3x(x - 16)$
(C) $3(x - 4)(x - 4)$
(D) $(x + 16)(x - 16)$

This problem combines common factor and the difference of two perfect squares. First, factor out the common factor of 3. What remains inside the parentheses is the difference of two perfect squares.

$$3(x^2 - 16) = 3(x + 4)(x - 4)$$

The correct answer is (A).

Quadratic Trinomials

A quadratic trinomial is an expression that contains a squared x-term, an x-term with a coefficient greater than zero, and a constant. These expressions can usually be factored into two binomials. To factor this type of expression, you need to find a combination of factors for the constant that when subtracted or added will result in the coefficient of the x-term. This is also a matter of using the correct sign in the factored binomial.

♦ Write $x^2 + 2x - 15$ in factored form.

The first term in each binomial will be x::

$$(x \quad)(x \quad)$$

The x times x will give us the x^2 term. In this example, we are looking for 2 numbers whose sum is +2 and whose product is -15, which means that one number is positive and one number is negative. The factors of 15 are: 1 & 15, or 3 & 5. Since the sum of the two numbers must be positive 2, the factors we are looking for are + 5 and - 3.

$(x + 5)(x - 3) = x^2 + 2x - 15$

To check the answer, use the FOIL (First Outside Inside Last) method of multiplying binomials.

Sometimes you will need to use factoring to obtain an answer, although the question is not looking for the complete factored form of the expression.

♦ Which is not a factor of $2x^2 - 20x + 48$?

(A) 2
(B) $x - 4$
(C) $x - 6$
(D) $x + 4$

First, always check for a common factor. In this case, the common factor is 2. We can rewrite the binomial as $2(x^2 - 10x + 24)$.

To factor the trinomial inside the parentheses, we are looking for two numbers whose product is 24 and whose sum is -10. Since the product is positive and the sum is negative, we know that both numbers must be negative. The factors of 24 are 1 & 24, 2 & 12, 3 & 8, and 4 & 6.

- 4 and - 6 are the combination we are looking for. The factored form is:

$$2(x - 4)(x - 6).$$

Therefore, (A) , (B) , and (C) are factors, making the correct answer (D).

FUNCTIONS

Two quantities, or two variables, can be related in many ways. When one quantity depends on a second quantity for its value, it is said to be a function of the second. If y depends on x, it is written y = f(x). "y" is the dependent variable, and x is the independent variable. For a car traveling at a constant speed, the distance it covers is a function of the time it travels. Algebraically, d= f(t).

Function notation may be written as h(x), g(x), f(t), etc. Functions can be represented as an equation, a graph, a mapping diagram, a table, or as a set of ordered pairs. Common notation for ordered pairs is (x,y). For a relationship to be a function there can be only one value of y for each value of x. When a function is graphed in the coordinate plane, the independent variable is placed on the horizontal axis and the dependent variable on the vertical axis. The "test" for determining whether a relationship is a function is the "straight line vertical test." If a vertical line is drawn through the graph, it can cross the graph in only one place.

The **domain** of a function is all the values that x, or the independent variable, can have (the replacement set of values for the independent variable). The **range** of the function is all the values that y, or the dependent variable, can have (the replacement set of values for the dependent variable, or of the function). The domain and range of each function is most easily seen from the coordinate graph of the function.

Evaluating functions

To find the f(3) for a function f(x) = 5x + 7 means to substitute 3 for x in the function. For f(x) = 5x + 7, f(3) = 5(3) + 7 = 15 + 7 = 22.

$f(x) = 3x - 2$
$f(-2) = 3(-2) - 2 = -6 - 2 = -8$

$g(x) = x^2 - 4$
$g(-2) = (-2)^2 - 4 = 4 - 4 = 0$

$h(x) = 2x^2 - 5x + 2$
$h(-2) = 2(-2)^2 - 5(-2) + 2 = 2(4) - 5(-2) + 2 = 8 + 10 + 2 = 20$

Arithmetic operations and composition with functions

There may be some problems on the test that combine the individual functions as discussed above into arithmetic operations with functions. These include:

sum of f and g	$(f + g)x = f(x) + g(x)$
difference of f and g	$(f-g)x = f(x) - g(x)$
multiplication of f times g	$(f * g)x = f(x) * g(x)$
division of f and g	$(f/g)x = f(x) / g(x)$

There may also be composite functions, which are written as f(g(x)), and is read "the f of the g of x". If $f(x) = 2x + 1$, and $g(x) = 2x^2 - 5$, then f(g(3)) means to find the value of the g(3) and use that value for x in the f(x) function. So in this case, it would be $g(3) = 2(3)^2 - 5 = 18 - 5 = 13$, then $f(13) = 2(13) + 1 = 27$.

Note that the g(f(3)) is NOT the same as the f(g(3)). Using the above functions, the g(f(3)) would be $f(3) = 2(3) + 1 = 7$, then $g(7) = 2(7)^2 - 5 = 98 - 5 = 93$.

This could all be done in one equation, if you do the algebraic conversion first. In the first example, f(g(x)) would mean that for every x in the f(x) equation, you would substitute the entire g(x) function;

$$2(2x^2 - 5) + 1 = 4x^2 - 10 - 1 = 4x^2 - 9$$

$$4(3)^2 - 9 = 4(9) - 9 = 36 - 9 = 27$$

PRACTICE — FUNCTIONS

1. What are the domain and range of this function?

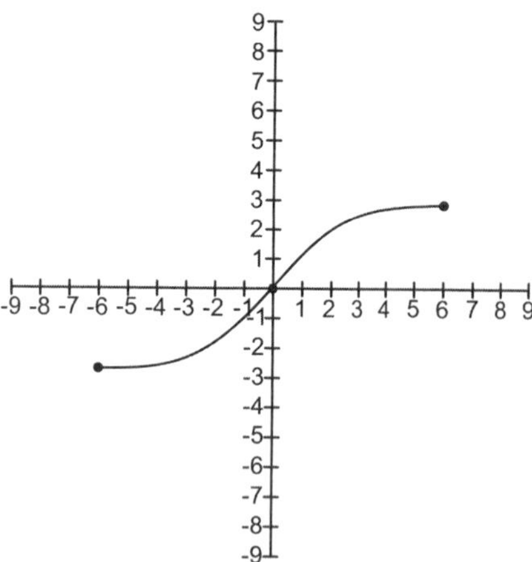

A) D:[-∞,∞] R:[-∞,∞]
B) D:[-3,3] R:[-6,6]
C) D:[-6,6] R:[-3,3]
D) D:[-6,∞] R:[-3,∞]

2. What are the domain and range of this function?

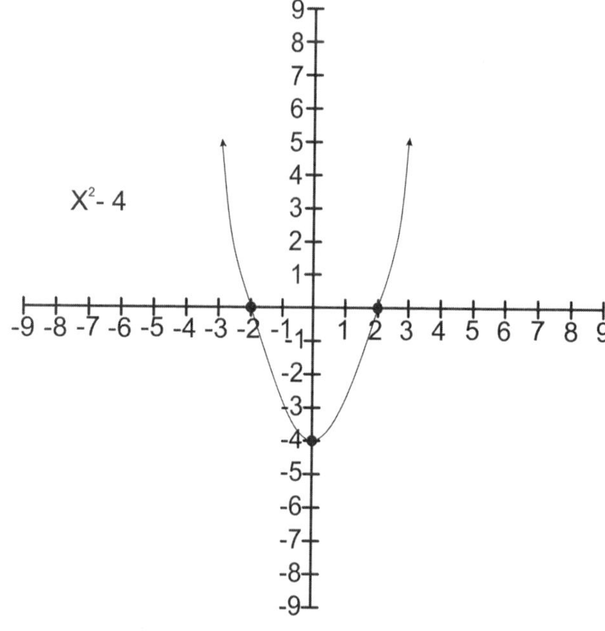

A) D:[-2,2] R:[-4,∞]
B) D:[-∞,∞] R:[-4,0]
C) D:[-∞,∞] R:[-4,∞]
D) D:[-4,∞] R:[-2,2]

3. Evaluate $f(x)=15x^2 + 9x - 6$ for f(2)

A) 30
B) 72
C) 42
D) 78

4. Which of the following is a function of x?

A.

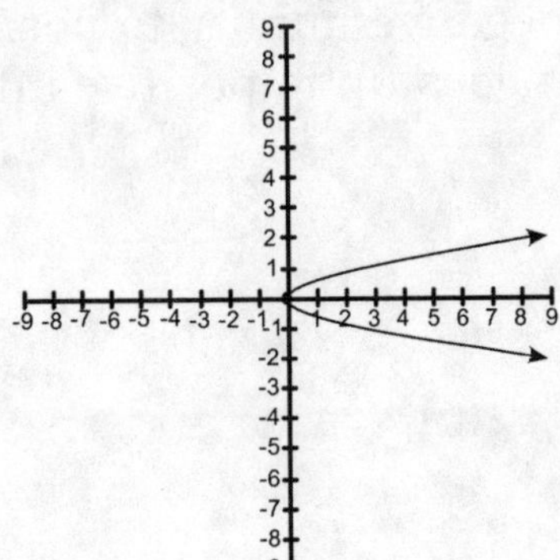

C.

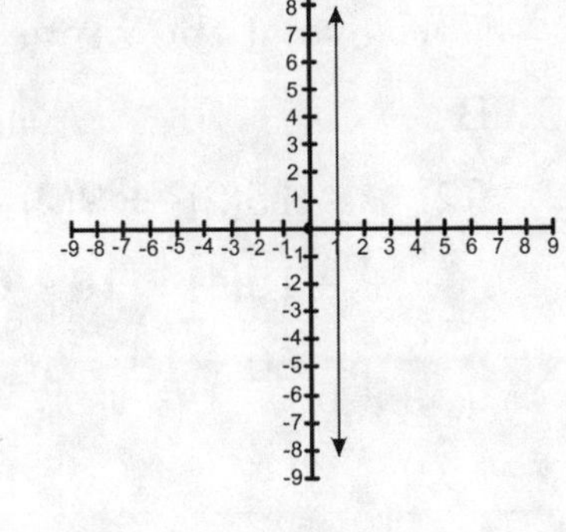

B.

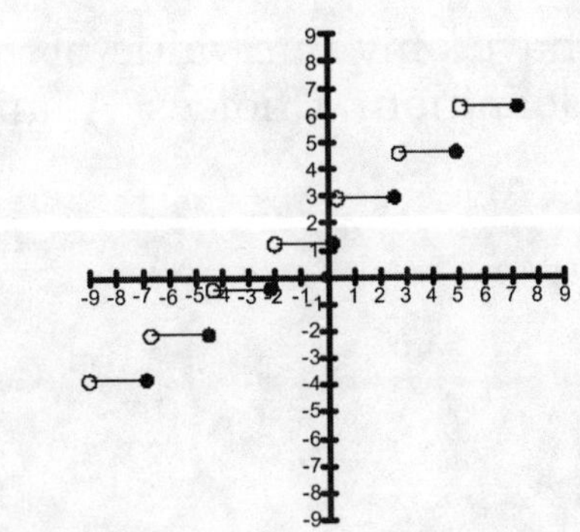

D.

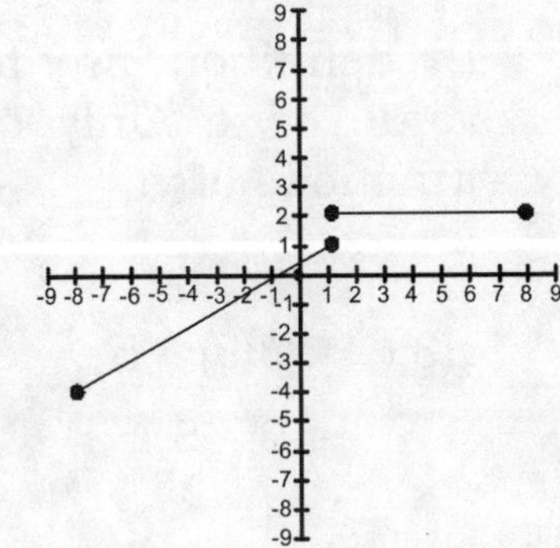

5. If $f(x) = x^2 + 5x + 6$ and $g(x) = 3x - 5$, what is f[g(3)] ?

A. 85
B. 42
C. 53
D. 80

ANSWERS — FUNCTIONS

1. C

   Domain is the scope of values of x. The line goes from -6 to 6. Range is the scope of the values of y. The line goes from -3 to 3.

2. C

   The domain can be any number from negative infinity to positive infinity. The lowest the range can be is -4 and it goes up to positive infinity.

3. B

   $f(2) = 15(2)^2 + 9(2) - 6$

   $= 15(4) + 18 - 6$

   $= 60 + 18 - 6$

   $= 72$

4. C

   By definition, any function of x would have only one value of y for each value of x. Only choice C fits that definition. Choices A and B are functions of y.

5. B

   $g(3) = 3(3) - 5$

   $= 9 - 5$

   $= 4$

   $f(g(3)) = f(4)$

   $= (4)^2 + 5(4) + 6$

   $= 16 + 20 + 6$

   $= 42$

Algebraic Equations

An algebraic equation is defined by two expressions separated by an equal sign. In order to be a valid equation, each side must actually be equal to the other. Both sides do not have to be algebraic expressions; one side is often a constant. To solve an algebraic equation, you must determine what value of the variable makes the expression valid. The most important thing to remember is that whatever you do to one side of the equation MUST be done to the other to maintain equality.

Fractional equations are equations where one or more terms are fractions.

To solve, find the common denominator of all the denominators. Then multiply each term by the common denominator. This will eliminate the fractions and you can solve like a simple equation.

♦ Solve: $\frac{3x}{4}+1=\frac{x}{3}+6$

The common denominator of 4 and 3 is 12.

$$12\left(\frac{3x}{4}+1\right) = 12\left(\frac{x}{3}+6\right)$$

$${}^{3}\cancel{12}\cdot\frac{3x}{\cancel{4}_{1}}+12\cdot 1 = {}^{4}\cancel{12}\cdot\frac{x}{\cancel{3}_{1}}+12\cdot 6$$

$$9x + 12 = 4x + 72$$

$$5x = 60$$

$$x = 12$$

Decimal equations can be solved by either multiplying by powers of 10 to eliminate the decimals or by combining the decimal parts. To avoid decimal problems, it is usually best to multiply by powers of 10.

♦ $.62x + 8.5 = .3x + 8.82$

Multiply by 100 to remove the decimals:

$$100(.62x) + 100(8.5) = 100(.3x) + 100(8.82)$$

$$62x + 850 = 30x + 882$$

$$32x = 32$$

$$x = 1$$

Most of the questions on a CLEP exam will combined a variety of skills. Below are examples of solving equations using absolute value, square roots, and factoring.

♦ What is the solution set for $\left| \frac{2}{3}x + 5 \right| = 9$?

(A) { 6 }
(B) { 21 }
(C) { 6, 21 }
(D) { 6, -21}

The absolute value will be 9 if the value of the algebraic expression is either +9 or -9. So there are two equations to solve:

$\frac{2}{3}x + 5 = 9$ **and** $\frac{2}{3}x + 5 = -9$

$\frac{2}{3}x = 4$ $\qquad \frac{2}{3}x = -14$

$x = 6$ **and** $x = -21$

The correct answer is (D).

♦ If $\sqrt{x+2}-3=5$, what does x equal?

To solve this equation, isolate the radical on one side and square the whole equation in order to eliminate the radical symbol.

$$\sqrt{x+2}-3=5$$

$$\sqrt{x+2}-3+3=5+3 \quad \text{add 3 to both sides of the equation}$$

$$\sqrt{x+2}=8$$

$$x+2=64 \quad \text{square both sides of the equation}$$

$$x+2-2=64-2 \quad \text{subtract 2 from both sides of the equation}$$

$$x=62$$

♦ If $x^2 + x - 12 = 0$, then x =

(A) -4 or -3
(B) -4 or 3
(C) 4 or -3
(D) 4 or 3

First, we factor the left side of the equation:

$(x + 4)(x - 3) = 0$

If the product is equal to zero, then one of the factors must be zero. So, either:

$x + 4 = 0$ **or** $x - 3 = 0$
$x = -4$ **or** $x = 3$

The correct answer is (B).

♦ What is the solution set for the equation $x^2 - 2x = 35$.

(A) { 0, 2 }
(B) { 35, 37 }
(C) { 7, - 5 }
(D) { -7, 5 }

The first thing we do is subtract 35 from both sides to form a quadratic equation equal to zero. Then we can proceed with factoring.

$x^2 - 2x - 35 = 0$

Factoring the left-hand side we get:

$(x - 7)(x + 5) = 0$

$x - 7 = 0$ or $x + 5 = 0$

$x = 7$ or $x = -5$

The correct answer is (C).

Simultaneous Systems of Equations

Systems of equations are most often found in word problems where two variables are found. If you express the problem with two equations using the two variables, you can solve the problem with the simultaneous equations. The first step is to create the equations from the word problems. Next, you must solve the system using substitution or combination.

The combination method should be used when one of the variables can be easily eliminated. For example, when one equation has a positive variable and the other equation has the same variable, but negative. Remember, you can multiply an equation by a constant and maintain equality, as long as you multiply every piece in the equation.

That way, if you have a variable that is has a positive coefficient of 2, and the other equation has that variable with a negative coefficient of 1, you should multiply the second equation by 2, thereby creating the situation where that variable can be eliminated through the combination method.

The substitution method involves solving one equation to get one of the variables expressed in terms of the other, then replacing that variable in the second equation. You will have one equation/one variable to solve.

♦ If $3x + y = 18$ and $x - y = 2$, solve for x and y. This is an example of a set of equations that has a variable that can be easily eliminated (y and –y).

$$\begin{array}{l} 3x + y = 18 \\ \underline{\;x - y = 2\;} \\ 4x \quad = 20 \\ \;x \quad = 5 \end{array}$$

Then:

$$5 - y = 2$$
$$y = 3$$

Remember to go back and substitute the value of 'x' into one of the equations to get the value of 'y'. To verify your work, use those values in both equations.

♦ $3x + 2y = 10$ and $2x - 3y = 11$

In this case, both equations must be multiplied to get a common coefficient for 'y'. Multiply the first equation by 3, the second by 2, then combine:

$$\begin{array}{rcl} 9x + 6y & = & 30 \\ + \underline{\;4x - 6y} & \underline{=} & \underline{22} \\ 13x & = & 52 \\ x & = & 4 \\ 3(4) + 2y & = & 10 \quad \text{Find y by substituting 4 for x.} \\ 12 + 2y & = & 10 \\ 2y & = & -2 \\ y & = & -1 \end{array}$$

Now try a word problem where you need to first determine the equations, then solve for the variables.

♦ A man paid \$39.00 for 5 shirts and 4 hats. At the same price, 4 shirts and 5 hats cost \$42.00. What is the price of each?

Let x = price of hats and y = price of shirts.

Therefore: $4x + 5y = 39$, and

$5x + 4y = 42$

Multiply each to get a common coefficient for x:

$$5(4x) + 5(5y) = 5(39)$$
$$4(5x) + 4(4y) = 4(42)$$

Since both coefficients are positive, subtract the equations:

$$\begin{array}{rcl} 20x + 25y &=& 195 \\ -\ \underline{20x + 16y} &=& \underline{168} \\ 9y &=& 27 \\ y &=& 3 \end{array}$$

Therefore, the shirts cost \$3.00 each. To find the price of hats, substitute:

$$\begin{array}{rcl} 4x + 5(3) &=& 39 \\ 4x + 15 &=& 39 \\ 4x &=& 24 \\ x &=& 6 \end{array}$$

Therefore, the hats cost \$6.00 each.

Quadratic Formula

It is possible to solve quadratic equations by using the quadratic formula, but it is usually more time-consuming. If the equation cannot be factored, then you may need to use the quadratic formula. The formula says the roots of $ax^2 + bx + c = 0$ are:

quadratic formula: $$x = \frac{-b \pm \sqrt{b^2 - 4ac}}{2a}$$

where a is the coefficient of the quadratic term, b is the coefficient of the linear term, and c is the constant.

♦ Solve for x: $x^2 - 3x = 40$.

Set the equation equal to zero first: $x^2 - 3x - 40 = 0$. Therefore, a =1, b = -3, c = -40.

$$x = \frac{-(-3) \pm \sqrt{(-3)^2 - 4(1)(-40)}}{2(1)}$$

$$x = \frac{3 \pm \sqrt{9+160}}{2}$$

$$x = \frac{3 \pm \sqrt{169}}{2}$$

$$x = \frac{3 \pm 13}{2}$$

$$x = \frac{3+13}{2} = \frac{16}{2} = 8 \quad \text{and} \quad x = \frac{3-13}{2} = \frac{-10}{2} = -5$$

Sometimes the roots (answers) are not whole numbers so you estimate.

♦ $x^2 - 7x = 10$

Set = 0: $x^2 - 7x - 10 = 0$

Therefore: a = 1, b = -7, c = -10

$$x = \frac{-(-7) \pm \sqrt{(-7)^2 - 4(1)(-10)}}{2(1)}$$

$$x = \frac{7 \pm \sqrt{49+40}}{2}$$

$$x = \frac{7+\sqrt{89}}{2}, \frac{7-\sqrt{89}}{2}$$ ($\sqrt{89}$ is approximately 9.4)

$$x = \frac{7+9.4}{2}, \frac{7-9.4}{2} = 8.2, -1.2 \text{ (approximate)}$$

PRACTICE — BASIC ALGEBRA AND FUNCTIONS

1. $3a - 5a + 7a =$

2. $(5a^2)(-2a^3) =$

3. $4c - 3a + 6c + a =$

4. $\frac{2a^2}{3b} \cdot \frac{18b^4}{6a^3} =$

5. Solve:

   A. $\frac{3ab}{2c} \div \frac{12b^2}{10c^2} =$

   B. $\left(\frac{2}{3}\right)^2 \cdot \left(\frac{1}{2}\right)^2 =$

6. Evaluate $3a^2 - 5a - 4$ if $a = -2$

7. If $f(x) = 2x + 3$, $f(0) =$

   (A) 0
   (B) 2
   (C) 3
   (D) x

8. If $f(x) = 2x^2 + 1$, then $f(x + 1) =$

   (A) $2x^2 + 2x + 1$
   (B) $2x^2 + 4x + 3$
   (C) $2x^3 + 4x + 2$
   (D) $2x + 3$

9. What is the next value for y in the table below?

x	2	3	4	5
y	6	12	20	?

   (A) 25
   (B) 40
   (C) 30
   (D) 35

10. If $f(x) = x^2$ and $g(y) = 2y$, $f[g(1)] =$

   (A) 4
   (B) 6
   (C) 2
   (D) 8

11. For $y = f(x) = 2x$, $x \in \{1, 2, ...10\}$, the first 3 elements of the range are:

(A) {1,2,3}
(B) {2,4,6}
(C) {∅}
(D) {8,9,10}

12. If $g(y) = y^2 + 2y + 1$, then $g(y - 1) =$

(A) $y^2 + 1$
(B) y^2
(C) $y^2 + 2$
(D) $y^2 + 2y$

13. What is the domain of {(1, 2), (2, 0), (3, 5)}?

(A) {1, 2, 3}
(B) {0, 2, 5}
(C) {1, 2}
(D) {2, 0, 5}

14. Factor:

A. $a^2 - 7a - 18$

B. $3a^2 - 11a + 6$

15. Simplify:

A. $(3x - 5)(2x + 4)$

B. $3(x + y) - (3x - y)$

16. Solve for x:

A. $3x - 7 = 2(x - 5)$

B. $3(x - y) = 2ab$

17. Solve for x:

$x = y + 2$

$2x - 5y = 1$

18. Solve for a:

$a + b = 4$

$a - b = c$

19. Solve for x:

$x^2 - 4x = 21$

20. Add: $\frac{3a}{x} + \frac{4}{y} =$

21. Solve for x: $\frac{3}{7} = \frac{12}{x}$

22. Solve for x: $3.5x + .7 = 7.7$

23. Solve for y: $y^2 - y = 12$

24. $|3a - 1| = 11$

25. Solve using the quadratic formula:

$y^2 - 3y = 5$

ANSWERS – BASIC ALGEBRA AND FUNCTIONS

1. $3a - 5a + 7a = 5a$

2. $(5a^2)(-2a^3) = 5 \times -2 \times a^2 \times a^3 = -10a^5$

3. $4c - 3a + 6c + a = 10c - 2a$

4. $$\frac{^{1}\cancel{2a^2}}{_{1}\cancel{3b}} \cdot \frac{\cancel{18b^4}^{\,6b^3}}{\cancel{6a^3}_{\,3a}} = \frac{1}{1} \cdot \frac{6b^3}{3a} = \frac{6b^3}{3a} = \frac{2b^3}{a}$$

5. A. $$\frac{3ab}{2c} \div \frac{12b^2}{10c^2} = \frac{^{1a}\cancel{3ab}}{_{1}\cancel{2c}} \cdot \frac{\cancel{10c^2}^{\,5c}}{\cancel{12b^2}_{\,4b}} = \frac{1a}{1} \cdot \frac{5c}{4b} = \frac{5ac}{4b}$$

   B. $$\left(\frac{2}{3}\right)^2 \cdot \left(\frac{1}{2}\right)^2 = \frac{^{1}\cancel{4}}{9} \cdot \frac{1}{\cancel{4}_{1}} = \frac{1}{9}$$

6. $$\begin{aligned} 3a^2 - 5a - 4 &= 3(-2)^2 - 5(-2) - 4 \\ &= 3(4) + 10 - 4 \\ &= 12 + 10 - 4 \\ &= 18 \end{aligned}$$

7. C $\quad f(0)=2(0)+3=3$

8. B $$\begin{aligned} f(x+1) &= 2(x+1)^2 + 1 \\ &= 2(x^2 + 2x + 1) + 1 \\ &= 2x2 + 4x + 2 + 1 \\ &= 2x^2 + 4x + 3 \end{aligned}$$

9. C By squaring each x and adding x to the product you get y. For example: $2^2 + 2 = 6$, $3^2 + 3 = 12$, and $4^2 + 4 = 20$. Therefore, $y = x^2 + x$, so $5^2 + 5 = 30$.

10. A $g(1) = 2(1) = 2$ and $f(2) = 2^2 = 4$.

11. B The range of a function is the set of y values of the ordered pairs (x, y). For the first three elements of the domain, {1,2,3}, y=2(1)=2, y=2(2)=4, and y=2(3)=6.

12. B $g(y-1) = (y-1)^2 + 2(y-1) + 1$

$= y^2 - 2y + 1 + 2y - 2 + 1$

$= y^2$

13. A The domain is the set of first numbers of the ordered pairs. In this set, the first elements of the ordered pairs are 1, 2, and 3.

14. A. $a^2 - 7a - 18 = (a - \quad)(a + \quad)$

$= (a - 9)(a + 2)$

B. $3a^2 - 11a + 6 = (3a - 2)(a - 3)$

15. A. $(3x - 5)(2x + 4) =$

$= (3x)(2x) + (3x)(4) + (-5)(2x) + (-5)(4)$

$= 6x^2 + 12x - 10x - 20$

$= 6x^2 + 2x - 20$

B. $3(x + y) - (3x - y) = 3x + 3y - 3x + y = 4y$

16. A. $3x - 7 = 2(x - 5)$ Proof: $3(-3) - 7 = 2(-3 - 5)$

$3x - 7 = 2x - 10$ $\quad -9 - 7 = 2(-8)$

$x = -3$ $\quad -16 = -16$

B. $3(x - y) = 2ab$ *or* $3(x - y) = 2ab$

$x - y = \frac{2ab}{3}$ $\quad 3x - 3y = 2ab$

$x = \frac{2ab}{3} + y$ $\quad 3x = 2ab + 3y$

$x = \frac{2ab + 3y}{3}$ $\quad x = \frac{2ab + 3y}{3}$

17. $x = y + 2$ and $2x - 5y = 1$, by substitution:

$2(y + 2) - 5y = 1$

$2y + 4 - 5y = 1$

$-3y + 4 = 1$

$-3y = -3$

$y = 1$

Therefore: $x = y + 2$

$x = 1 + 2$

$x = 3$

18. $a + b = 4$

$\underline{a - b = c}$

$2a = 4 + c$

$a = \frac{4 + c}{2}$

19. $x^2 - 4x = 21$ change to: $x^2 - 4x - 21 = 0$

$(x - 7)(x + 3) = 0$

$x - 7 = 0 \quad x + 3 = 0$

$x = 7 \quad x = -3$

20. $\frac{3a}{x}+\frac{4}{y}=\frac{3a\cdot y+4\cdot x}{xy}=\frac{3ay+4x}{xy}$

21. $\frac{3}{7}=\frac{12}{x}$

$3x = 84$

$x=\frac{84}{3}=28$

22. $3.5x + .7 = 7.7$

$35x + 7 = 77$

$35x = 70$

$x = 2$

Proof: $3.5(2) + .7 = 7.7$

$7 + .7 = 7.7$

$7.7 = 7.7$

23. $y^2 - y = 12$

1. $y^2 - y - 12 = 0$
2. $(y - 4)(y + 3) = 0$
3. $y - 4 = 0 \qquad y + 3 = 0$
4. $y = 4 \qquad y = -3$

Proof: $4^2 - 4 = 12$ $\quad (-3)^2 - (-3) = 12$

$16 - 4 = 12 \quad 9 + 3 = 12$

$12 = 12 \quad 12 = 12$

24. $+(3a - 1) = 11$

$3a - 1 = 11$

$3a = 12$

$a = 4$

$-(3a - 1) = 11$

$-3a + 1 = 11$

$-3a = 10$

$a = -\frac{10}{3} = -3\frac{1}{3}$

25. $y^2 - 3y - 5 = 0$

$a = 1,\ b = -3,\ c = -5$

$$x=\frac{-(-3)\pm\sqrt{(-3)^2-4(1)(-5)}}{2(1)}$$

$$x=\frac{3\pm\sqrt{9+20}}{2}$$

$$x = \frac{3+\sqrt{29}}{2}, \frac{3-\sqrt{29}}{2}$$

$$x = \frac{3+5.38}{2}, \frac{3-5.38}{2}$$

x = 4.19,   x = -1.19

WORD PROBLEMS

The word problems included in College Mathematics are grouped according to type: integer, age, digit, motion, work, and fraction.

Integer Problems

Many integer problems simply relate facts about two or more unknown integers. To solve, choose a variable to represent one of the integers and then use the facts given to relate the two unknowns.

> ♦ The sum of two integers is 263. The larger integer is 31 more than three times the smaller integer. Find both integers.

Let n =	the smaller integer
Then 263 – n =	the larger integer
263 – n = 3n + 31	Translating the information in the problem that states that the larger integer (263 – n) is 31 more than three times the smaller integer (n).
263 – n – 31 = 3n	Subtracting 31 from both sides.
232 = 4n	Adding n to both sides.
58 = n	Dividing both sides by 4. This gives us the smaller integer which we called "n" above.
263 – 58 = 205	This gives us the larger integer, plugging 58 into the 263 – n = larger integer equation.

So, the two integers are 58 and 205.

In other problems about integers, it is important to recognize the type of integers involved: consecutive integers such as 5, 6, 7; consecutive odd integers such as 5, 7, 9; and consecutive even integers such as 6, 8 10. Integers could also be negative, so thinking must include those such as -4, -3, -2 for consecutive integers, or -9, -7, -5 for consecutive odd integers. Consecutive even integers could include -2, 0, 2, 4.

To solve consecutive integer problems, the algebraic expressions to represent four consecutive integers would be:

Let x = the first integer.

Then $x + 1$ = second consecutive integer

$x + 2$ = third consecutive integer

$x + 3$ = fourth consecutive integer, and so on

To solve the problem, the relationship between the integers must be discovered from reading the problem. Often the sum of the integers is given, but not always. An integer problem may read:

> ♦ Find three consecutive integers such that the sum of the first and third is 9 more than the second.

Let x = the first integer

$x + 1$ = the second consecutive integer

$x + 2$ = the third consecutive integer

Then $x + (x + 2) = (x + 1) + 9$	Translating the information in the problem
$2x + 2 = x + 10$	Combining like terms
$x = 8$	Solving the equation.

Therefore, the consecutive integers are 8, 9 and 10.

To check your answer:

the sum of the first and third integers, 8 and 10, is 18, which is 9 more than the second integer, 9.

To solve integer problems involving either consecutive odd or consecutive even integers, the algebraic expressions to represent either would be:

Let x = the first consecutive odd (even) integer

Then $x + 2$ = the second consecutive odd (even) integer

$x + 4$ = the third consecutive odd (even) integer

Whether you start with an even or an odd integer, the next one will always be two more than that first one and the one after that will be 4 more than the integer with which you started.

Age Problems

Age problems usually connect the ages of two or more people, often as they are related now, and how the relationship changes in time (or has changed from some time ago). To solve, choose a variable to represent the age of one person now, and express the other age(s) in terms of that variable. Then, using the information given, express the ages in future time (or in the past) and write an equation connecting the information given.

Billy is three times Sally's age today. Five years ago, he was five times her age. How old are Billy and Sally today?

To solve this problem, you have to assign a variable to Sally's age:

x = Sally's age today

Therefore, you can say that Billy's age today is $3x$ (three times Sally's age today).

$3x$ = Billy's age today

The next step is to translate the second sentence into an equation, using the variables established above:

Billy's age 5 years ago = $3x - 5$

Sally's age 5 years ago = $x-5$

The statement says that 5 years ago, Billy was 5 times Sally's age:

$$3x - 5 = 5(x - 5)$$

Now solve for x:

$$3x - 5 = 5x - 25$$

$$-2x = -20$$

$$x = 10$$

Inserting x into the initial variable assignments gives the answer:

Sally's age today = 10

Billy's age today = 30

Digit Problems

The digits that make up a number may be represented by letting h = the hundreds digit, t = the tens digit, and u = the units digit. A three-digit number would be represented as $100h + 10t + u$. A two-digit number would be represented as $10t + u$. When the problem involves a two-digit number reversed, it would be represented as $10u + t$.

♦ Find a two digit number such that when the original number is reversed, the new number is 23 less than two times the original number. The sum of the digits is 14.

Let t = tens digit of the original number

u = units digit of the original number

Then $10t + u$ = original number

From the information given in the problem:

u = tens digit of the number reversed

t = units digit of the number reversed

Then $10u + t$ = the number reversed.

$t + u = 14$	from the information in the problem
$10u + t = 2(10t + u) - 23$	translating that the new number (number with digits reversed) is 23 less than two times the original number $(10t + u)$
$10u + t = 20t + 2u - 23$	Distributing the 2
$8u + t = 20t - 23$	Subtracting 2u from both sides
$8u - 19t = -23$	Subtracting 20t from both sides

At this point, you have two equations and two variables:

$$u + t = 14$$

$$8u - 19t = -23$$

One way to solve this system of equations is to multiply the top equation by -8:

$$\begin{array}{rcr} -8u & - & 8t = -112 \\ \underline{8u} & \underline{-} & \underline{19t = -23} \\ & & -27t = -135 \\ & & t = 5 \end{array}$$

and therefore, u = 9 (from the original equation t + u = 14)

The final answer to the question is 59.

Motion Problems

Motion problems are usually one of three types: round trip, motion in opposite directions, or motion in the same direction (usually a "catch-up" problem). The formula to remember for all motion problems is distance = rate * time, D=RT. It will often be necessary to use it in the form R=D/T, or T=D/R, depending on the information given and how it is connected in the problem.

If you know any two of the three elements, distance, rate or time, you can solve for the other. Many motion problems have two vehicles traveling toward or away from each other. In many cases, the distance, rate or time of the two is the same.

> ♦ Two cars are traveling toward each other at 50 MPH and 40 MPH respectively. If they leave at the same time from a distance of 270 miles apart, when will they meet?

In this problem, the cars travel the same amount of time.

Since $D = R \times T$, then $T = \dfrac{D}{R}$.

The first car's time $= \dfrac{\text{x miles}}{\text{50 MPH}}$, and the second car's time $= \dfrac{270 - x}{\text{40 MPH}}$.

(Note: 270 - x is the distance remaining).

Since the time traveled is the same: $\dfrac{x}{50} = \dfrac{270 - x}{40}$.

$$\begin{array}{lrcl} \text{Cross multiplying:} & 40x & = & 50(270 - x) \\ \text{Dividing by 10:} & 4x & = & 5(270 - x) \\ & 4x & = & 1350 - 5x \\ & 9x & = & 1350 \\ & x & = & 150 \text{ miles} \end{array}$$

The first car travels a distance of 150 miles at 50 MPH. Therefore, the time will be $\frac{150}{50} = 3$ hours.

Another way to solve this problem is to figure out how fast the cars are traveling toward each other: 40 + 50 MPH. Divide this result into the total distance traveled: $\frac{270}{90} = 3$ hours.

A useful ratio in motion problems is 30 MPH = 44' per second. This ratio is useful with problems that require you to change MPH to feet per second.

Example: A plane travels at 600 MPH. How many feet per second is this?

Solution: $\frac{30 \text{ MPH}}{44 \text{ FPS}} = \frac{600 \text{ MPH}}{x \text{ FPS}}$

$$30x = 600 \times 44$$
$$x = 20 \times 44$$
$$x = 880 \text{ feet per second}$$

Work Problems

Work problems are often found on the exam. The key to work problems is:

1. Express the rate of work done each day (or hour) as a fraction.
2. Let x represent the time required for each of two or more working together to complete the job.
3. Let the completed job be represented by 1.

♦ If John does a job in 6 hours, Bob does the same job in 3 hours; how long will it take both, working together?

John does $\frac{1}{6}x$ per hour. Bob does $\frac{1}{3}x$ per hour.

Together: $\frac{1}{6}x + \frac{1}{3}x = 1$

Multiply both sides by 6 to clear the fractions and solve:

$$6 \cdot \frac{1}{6}x + 6 \cdot \frac{1}{3}x = 6 \cdot 1$$
$$1x + 2x = 6$$
$$3x = 6; \quad x = 2 \text{ hours}$$

Fraction Problems

To solve fraction word problems, represent the numerator or denominator with an unknown. Multiply the equation by the common denominator of the denominators.

♦ A fraction equal to $\frac{3}{5}$ has the numerator increased by 4 and denominator doubled. The new fraction is $\frac{1}{2}$. Find the original fraction.

Since the original fraction is equal to $\frac{3}{5}$, it can be written $\frac{3x}{5x}$.

$$\text{Therefore: } \frac{3x+4}{2(5x)} = \frac{1}{2}$$

$$2(3x + 4) = 10x$$

$$6x + 8 = 10x$$

$$8 = 4x$$

$$2 = x$$

Since $x = 2$: $\frac{3x}{5x} = \frac{3(2)}{5(2)} = \frac{6}{10}$ was the original fraction.

PRACTICE — WORD PROBLEMS

1. A merchant sells two blends of coffee. In the first, 2 pounds of \$0.60 coffee are mixed with one pound of \$0.90 coffee. In the second, 2 pounds of \$0.90 coffee are mixed with one pound of \$0.60 coffee. What is his additional profit if he sells 30 pounds of the first blend at the average price per pound of the second blend?
2. A man drives 100 miles one way on a trip at an average speed of 30 MPH. His return rate was only an average of 20 MPH. What is his average rate for the entire trip?
3. A pool has 2 inlet pipes. Each pipe used alone can fill the entire pool in 12 hours. What part of the pool can be filled in 3 hours if both pipes are used together?
4. John has a coin collection of 20 coins. The number of half-dollars is 2 more than the number of quarters, and the number of dimes is 6 more than the number of quarters. What is the value of the collection?
5. A missile goes 5,400 miles in 45 minutes. How much farther will it go in another 50 minutes at the same speed?
6. If John takes 30 minutes to get to work 20 miles away, how much faster must he travel if he leaves 10 minutes late?
7. A total of \$7500 is invested at two different rates, 4% and 5%. How much money is invested at each rate if the interest on the amount invested at 4% is \$75 more than the amount at 5%?
8. Find two numbers such that the larger is 3 times the smaller, and the sum of the larger and twice the smaller is 45.
9. Find four consecutive odd integers such that the sum of the first 3 integers is six more than the last.
10. In a football league each team plays every other team. If there are 28 games played, how many teams are in the league?

ANSWERS – WORD PROBLEMS

1. In the first blend the value is 2(.60) + .90 = \$2.10. The second is worth 2(.90) + .60 = \$2.40. The average price of the first is $\frac{2.10}{3}$ = \$0.70 per lb. The average price of the second is $\frac{2.40}{3}$ = \$0.80 per lb. If the first mixture is sold at the average price of the second, he will make \$0.10 per lb. extra. For 30 pounds, therefore, he will make 30 · (.10) = \$3.00 profit.

2. For this problem, you must remember that an average is a total divided by a total. In this case, to obtain the average rate for the whole trip, you must divide the total distance traveled (200 miles) by the total time traveled.

   Time traveled on the way there: $\frac{100}{30} = 3\frac{1}{3}$ hours

   Time traveled on the way home: $\frac{100}{20} = 5$ hours

   Total time traveled: $8\frac{1}{3}$ hours

   The average rate would be:

$$\frac{200}{8\frac{1}{3}} = \frac{200}{\frac{25}{3}} = 200 \text{ x } \frac{3}{25} = \frac{600}{25} = 24 \text{ mph}$$

3. Letting the amount done by each pipe be represented by a fraction can solve this problem.

   Therefore: $\frac{x}{12} + \frac{x}{12} = 1$

$$x + x = 12$$

$$2x = 12$$

$$x = 6$$

   Therefore both pipes together can fill the pool in 6 hours. In 3 hours $\frac{3}{6} = \frac{1}{2}$ of the pool will be filled.

4. Let x represent the number of quarters, x + 2 represent the number of half dollars, and x + 6 represent the number of dimes.

Therefore: $x + (x + 2) + (x + 6) = 20$

$$3x + 8 = 20$$

$$3x = 12$$

$$x = 4$$

and $4(.25) + 6(.50) + 10(.10) = \$1.00 + \$3.00 + \$1.00 = \$5.00$

5.
$$D = R \times T$$

$$5{,}400 = R \times 45 \text{ min.}$$

$$R = 120 \text{ mi./min.}$$

Therefore: $D = 50 \text{ min. } (120 \text{ miles/min.})$

$$D = 6{,}000 \text{ miles}$$

You can also solve by using ratio:

$$\frac{5{,}400}{R} = \frac{45}{50}.$$

6. In order to go 20 miles in 20 min. (10 min. late) John will have to travel 60 MPH (a mile a minute). His usual rate is $R = \frac{20}{.5} = 40$ MPH. Therefore he will have to go 20 MPH faster.

7. Let x represent the amount invested at 4%. Then (7500 - x) is the amount at 5%. Therefore: $4\%(x) = \$75 + 5\%(7500-x)$

$$4\%x = \$75 + 375 - 5\%x$$

$$9\%x = 450$$

$$x = \frac{450}{.09} = \$5000 \text{ at } 4\%$$

$$7500 - 5000 = 2500 \text{ at } 5\%$$

8. Let x = smaller number and 3x = larger number.

$$3x + 2(x) = 45$$

$$5x = 45$$

$$x = 9$$

$$3x = 27$$

9. x = 1st, $x + 2$ = 2nd, $x + 4$ = 3rd, $x + 6$ = 4th.

$$x + (x + 2) + (x + 4) = 6 + (x + 6)$$
$$3x + 6 = 6 + x + 6$$
$$2x = 6$$
$$x = 3$$

Therefore the answer is: 3, 5, 7, 9.

10. If each team plays every other team, let n = number of teams. Then n(n - 1) = total number of games (since no team plays itself). However, this total is twice as large since "A plays B" is the same as "B plays A".

Therefore: $$\frac{n(n-1)}{2} = 28$$
$$n(n - 1) = 56$$
$$n^2 - n - 56 = 0$$

Factor: $(n - 8)(n + 7) = 0$

$n - 8 = 0 \quad n + 7 = 0$

$n = 8 \quad n = -7$ (reject, can't be negative)

INEQUALITIES

An inequality is a statement in which two quantities are not equal. One may be greater than or less than another. Both sides of an inequality can be added to, or subtracted from, without changing the direction of the inequality. This is also true for multiplying and dividing by a positive number.

If $a > b$, then $a + c > b + c$

However, if an inequality is divided by a negative number, the direction of the inequality must be switched.

We can work with inequalities the same way we solve equations.

♦ If $3x + 2 > x + 10$ then,

$2x > 8$ (subtracting x and 2 from both sides)

$x > 4$ (dividing by 2)

Our answer says that $3x + 2$ is greater than $x + 10$ whenever x is greater than 4. (Try substituting a number greater than four.)

Remember, if an inequality has a negative coefficient, reverse the direction of the inequality sign in solving.

♦ If $-5x - 2 > -x + 18$

then $-4x > 20$

(dividing by -4) $x < -5$

This says that $-5x - 2$ is greater than $-x + 18$ whenever x is less than -5. Try substituting a number less than -5 to prove the inequality.

Some inequalities are presented as representing a range of numbers, and ranges can also be combined.

♦ If $3 < r < 9$ and $1 < s < 10$, then the

range: $(3 + 1) < r + s < (9 + 10)$

and $(3 - 10) < r - s < (9 - 1)$

$(3 \cdot 1) < r \cdot s < (9 \cdot 10)$

$(3 \div 10) < r \div s < (9 \div 1)$

If an inequality is presented and you are asked to compare parts, use one of the principles involving adding or multiplying.

♦ If y is positive and $\frac{x}{y} > 3$, which is greater, x or 2y?

If we multiply both sides of the inequality by y we get:

$$y \cdot \frac{x}{y} > 3$$

$$x > 3y$$

Therefore, x is greater than 2y.

PRACTICE PROBLEMS— INEQUALITIES

1. If a > b, are integers then which of the following is/are not necessarily true?

   A. $ac > bc$

   B. $a^2 > b^2$

   C. $a + c > b + c$

   D. $\frac{a}{c} > \frac{b}{c}$

2. Solve for x: $3x + 2 > 11$

3. Solve for x: $-3x + 2 > 11$

4. If $\frac{P}{N} > 2$, then P > ? (N is not zero.)

5. $4 < x < 10$ and $-2 < y < 12$. What is the range of $x + y$?

6. Which of the following could not be true if $a + a^2 < 1$?

   A. $a > 1$ B. $a > -1$ C. $a < 0$ D. $a > 0$

7. Solve for x: $3x - 8 < x - 18$

8. If $0 < a - b < 3$, then a ? b

9. If $a > 0$, which is greater, a^2 or a?

10. If $a < 0$, which is greater, a^2 or a?

11. If $a < 5 < b$ (a and b are positive) then which is true:

    A. $\frac{1}{a}<\frac{1}{5}<\frac{1}{b}$ B. $\frac{1}{b}<\frac{1}{5}<\frac{1}{a}$ C. $\frac{1}{a}<\frac{1}{b}<\frac{1}{5}$ D. $\frac{1}{b}<\frac{1}{a}<\frac{1}{5}$ E. $\frac{1}{b}<5<\frac{1}{5}$

12. Solve for y in terms of x: $3x + y > xy$.

ANSWERS

1. A. False—if a and b are positive and c is negative, then $ac < bc$

   B. False—if a and b are both negative, then $a^2 < b^2$

   C. True

   D. False—if a and b are positive and c is negative, then $\frac{a}{c} < \frac{b}{c}$

2. $3x + 2 > 11$

   $3x > 9$

   $x > 3$

3. $-3x + 2 > 11$

   $-3x > 9$

   $x < -3$

4. $\frac{P}{N} > 2$ Therefore, $P > 2N$ (when $N>0$) or $P < 2N$ (when $N<0$)

$N > 0$	$N < 0$
$\frac{P}{N} > 2$	$\frac{P}{N} < 2$
$P > 2N$	$P < 2N$

5. $+4 < x < +10$

   $-2 < y < +12$

   $\overline{+2 < x + y < +22}$

6. Choice A could not be true for any of the numbers in the range: If $a > 1$ and you add another positive number to it you will always have a number that is greater than 1.

7. $3x - 8 < x - 18$

   $2x < -10$

   $x < \frac{-10}{2}$

   $x < -5$

8. If $0 < a - b < 3$

adding b: $0 + b < a < 3 + b$

therefore: $b < a < 3 + b$

and then the "?" should be replaced with ">"; $(a > b)$

9. If $a > 1$, then $a^2 > a$. Example: $3^2 > 3$.

If $0 < a < 1$, then $a^2 < a$. Example: $\left(\frac{1}{2}\right)^2 < \frac{1}{2}$.

10. Since $a < 0$, a is a negative number. Squaring a negative number produces a positive. Therefore, $a^2 > a$.

11. The correct answer is B. If $a < 5 < b$, since a and b are positive the reciprocal will reverse the order of the inequality.

12. solution: $3x + y > xy$ or $xy < 3x + y$

$$y - xy > -3x \qquad xy - y < 3x$$

$$y(1 - x) > -3x \qquad y(x - 1) < 3x$$

$$y > \frac{-3x}{1-x} \qquad y < \frac{3x}{x-1}$$

As you can see, the two answers are equivalent.

SET THEORY

A number set is simply a group of numbers.

A = {3, 6, 9} is the set of all numbers between 1 and 10 that are multiples of 3.

B = {6} is a subset of A. It could represent all numbers between 1 and 10 that are not only multiples of 3, but also multiples of 6.

We use the symbol $B \subset A$ to denote the subset relationship.

The set of all odd integers evenly divisible by 2 could be represented as: { } or $\varnothing$. There are no members of this set. It is called the **null** set or **empty** set.

♦ How many subsets are there of set A, if A = {3, 6, 9}?

(A) 3
(B) 6
(C) 8
(D) 9

The set itself and the null set are considered subsets of the set.

If we just try to list them, we would have:

{3}, {6}, {9}, {3,6}, {3,9}, {6,9}, {3, 6, 9}, and { }, so there are 8 subsets and the correct answer is (C).

In general, the number of subsets is equal to 2^n, where n = the number of members in the set. In our example, n = 3 and thus $2^n = 2^3 = 8$ is the number of different subsets.

Set Relationships

We've already studied one relationship between sets, and that's the subset relationship, where all the members of the subset were also members of the original set. We want to discuss two other relationships, the union of two or more sets and the intersection of two or more sets.

The **union** of two sets is a set of all elements contained in either of the original sets. It is represented by the symbol $\cup$.

♦ If A = {x: x is even and 0 < x < 12} = {2, 4, 6, 8, 10}, and

B = {x: x is a perfect square and 0 < x < 11} = {1, 4, 9},

then the union of these two sets, represented as **A ∪ B** is all elements that are in either **A or B or both.**

$A \cup B$ = {1, 2, 4, 6, 8, 9, 10}.

The **intersection** of two sets is a set of all elements common to both sets. It is represented by the symbol $\cap$.

- If A = {positive odd integers < 30} and B = {perfect squares < 30}, then how many elements are there in $A \cup B$?

  (A) 17
  (B) 20
  (C) 3
  (D) 15

Listing the members of each set:

A = {1, 3, 5, 7, 9, 11, 13, 15, 17, 19, 21, 23, 25, 27, 29}

B = {1, 4, 9, 16, 25}

$A \cup B$ = {1, 3, 4, 5, 7, 9, 11, 13, 15, 16, 17, 19, 21, 23, 25, 27, 29}

The correct answer is (A).

By inspection, we know that there are 30 integers between 1 and 30 inclusive, and half of them are odd, or 15. The union would include any perfect squares that were even, or 4 and 16. 15 + 2 = 17.

- If A = {2, 4, 6, 8, 10}, B = {10, 15, 20, 25}, and C = {3, 6, 9, 12, 15}, then what is $A \cap (B \cup C)$?

  (A) {2, 4, 6, 8, 10, 15}
  (B) {10}
  (C) {6, 10}
  (D) {2, 3, 4, 6, 8, 9, 10, 12, 15, 20, 25}

As with any mathematical operation, perform the operation in the parentheses first:

{$B \cup C$} = {3, 6, 9, 10, 12, 15, 20, 25}, any element that is in one or both of the sets.

$A \cap (B \cup C)$ are the elements in common between A and the union of B and C. The common elements are 6 and 10.

The correct answer is (C).

The final relationship we want to discuss is **complement**. It deals with the relationship between 2 specific subsets of a given set. If B is a subset of A, the complement of B (written ~B) is a subset containing all the elements of A not included in B.

♦ If A = {all integers between 1 and 10 inclusive} and B = {all even integers between 1 and 10 inclusive}, then the complement of B is {all odd integers between 1 and 10 inclusive}.

Venn Diagrams

We can use circles to represent the relationship between sets. These circle representations are called Venn diagrams. For example:

A = {4, 8, 12, 16}

B = {6, 8, 12, 20}

C = {4, 12, 20, 28}

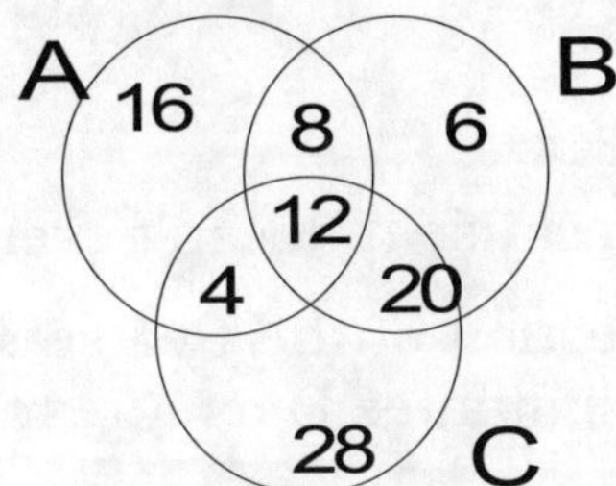

A ∩ B = {8, 12}

A ∩ C = {4, 12}

B ∩ C = {12, 20}

A ∩ B ∩ C = {12}

♦ In the Venn diagram below, the shaded region represents which relationship?

(A) A ∪ B
(B) A ∪ (B ∪ C)
(C) A ∪ C
(D) A ∪ (B ∩ C)

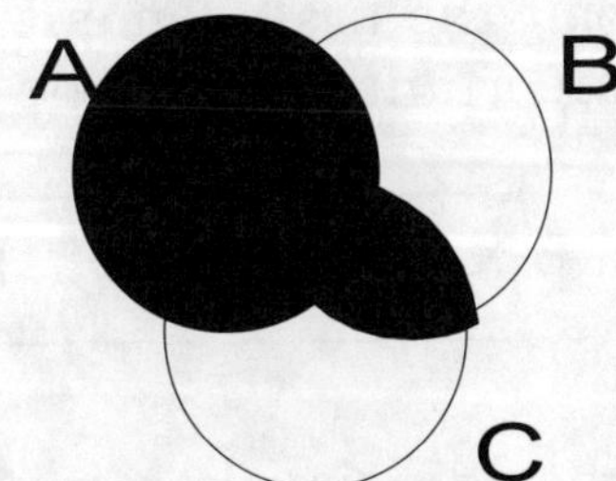

By inspection, we see that we've shaded all of circle A plus the area where circles B and C meet, or intersect. So, we are joining A with the intersection of B and C. The correct answer is (D).

♦ Given the Venn diagram, which of the following statements are true?

I. $A \cup B = \{4, 6, 8\}$

II. $A \cap C = \{5, 6, 7\}$

III. $B \cap C = \{16, 24\}$

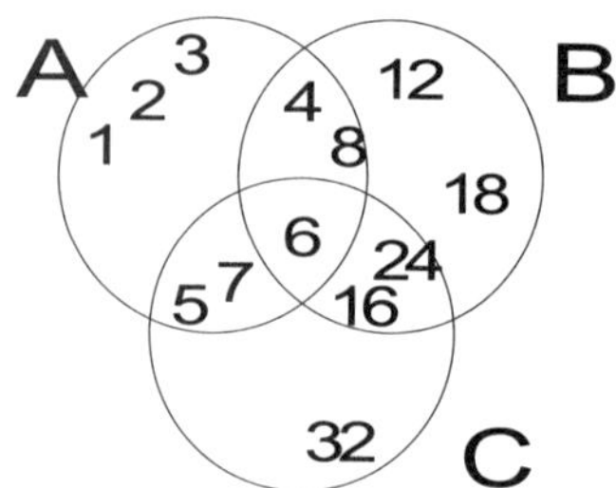

(A) I and II
(B) II only
(C) All
(D) None

A is obviously false because it doesn't even include all the elements of A.

{4, 6, 8} is the intersection of the two sets. Therefore, we can eliminate (A) and (C). II is true which eliminates D as a response. The correct answer is B.

The Cartesian product of two sets is the set of all ordered pairs that pairs each element from the first set with each element from the second set. The Cartesian product is shown by A X B.

♦ If $A = \{1, 2, 3\}$ and $B = \{a, b\}$,
then $A \times B = \{(1, a), (1, b), (2, a), (2, b), (3, a), (3, b)\}$.

Notice that in every pair the first element is from the first set, the second from the second set. The important point is the order.

♦ For the same sets, $B \times A = \{(a, 1), (b, 1), (a, 2), (b, 2), (a, 3), (b, 3)\}$.

These are different ordered pairs than in A X B. Can you see that $(A \times B) \cap (B \times A) = \varnothing$? Note that if A has 3 elements, and B has 2 elements, then A X B has $3 \times 2 = 6$ ordered pairs. This is a simple but important counting principal.

Closure

Closure is a property of sets with respect to an operation. For example, the set of all integers is **closed** with respect to addition. This means that the result of adding any two members of the set of integers is also a member of the set of integers. We know this is true because we know that adding two integers results in an integer. The set of all integers is also closed for multiplication.

Closure, a property of sets, regards the effect of a certain operation on the elements of a set. Closure asks for a given operation (such as addition, subtraction, multiplication, or division) performed on a given set if all results are within the given set. If any result is outside the set, the set is not closed. If all results are within the set, the set is closed.

♦ The set {1, 0} is closed under multiplication since all the answers are in the set: $1 \bullet 1 = 1$, $1 \bullet 0 = 0$, $0 \bullet 0 = 0$.

The operation (in this example, multiplication) can be performed on an element with itself. For this set, {1, 0}, the set is not closed under addition, since $1 + 1 = 2$ and 2 is not an element of the set.

♦ Is the set of even numbers closed under addition?

The answer is yes, since adding even numbers gives an even number which, therefore, must be in the given set. How about the set of odd numbers under addition? (no; adding two odd numbers gives an even number, which is not in the original set)

♦ $A = \{x: x \geq 1\}$, $B = \{x: x \leq 1\}$. Find $A \cap B$.

This is read: A is the set of numbers x, such that x is greater than or equal to 1. B is the set of numbers x, such that x is less than or equal to 1. Find the intersection of sets A and B.

A favorite type of set problem which uses set symbolism might be like this:

Since A is the set of numbers from 1 up and B is the set of numbers from 1 down, the only common element is 1. Therefore, $A \cap B = \{1\}$.

NOTE: Sometimes the symbol | is used in place of the colon to mean "such that." $A = \{x \mid x \geq 0\}$ says A is the set of elements x such that each element is greater than or equal to zero.

PRACTICE PROBLEMS — SET THEORY

Questions 1 - 7 refer to the following sets:

A = {1, 2, 3, 4, 5} B = {2, 3, 5, 8, 9} C = {2, 4, 6, 7, 8}

1. What is A ∪ B?

   (A) {2,3}
   (B) {2,3,5}
   (C) {1,4,8,9}
   (D) {1,2,3,4,5,8,9}

2. What is A ∩ B?

   (A) {1,2,3,4,5,8,9}
   (B) {∅}
   (C) {2,4}
   (D) {2,3,5}

3. What is A ∪ (B ∩ C)?

   (A) {2,3,5}
   (B) {1,2,3,4,5,8}
   (C) {2,3,4,5}
   (D) {2,3,4,5,6,7,8,9}

4. How many subsets of C are there?

   (A) 5
   (B) 7
   (C) 32
   (D) 16

5. Which of the following is **not** an element of B X C?

   (A) (3,4)
   (B) (2,6)
   (C) (2,8)
   (D) (3,9)

6. How many pairs are in the set A X C?

   (A) 5
   (B) 10
   (C) 25
   (D) 7

7. What is (A ∩ B) ∩ C?

   (A) {2}
   (B) {2,3,5}
   (C) {∅}
   (D) {2,3,4,5,6,8}

8. Which of the following sets is closed under addition?

(A) {0,1}
(B) {-1,0,1}
(C) {even numbers}
(D) {odd numbers}

9. If A = {x: $-1 < x \leqq 3$} and B = {X: $x \geqq 3$}, then $A \cup B$ =?

(A) $x < -1$
(B) $x > -1$
(C) $x \geqq 3$
(D) $x \geqq -1$

10. In the Venn diagram below, the shaded region represents which of the following?

(A) $A \cup B \cup C$
(B) $A \cap B \cap C$
(C) $(B \cup C) \cap A$
(D) $(A \cap C) \cup C$

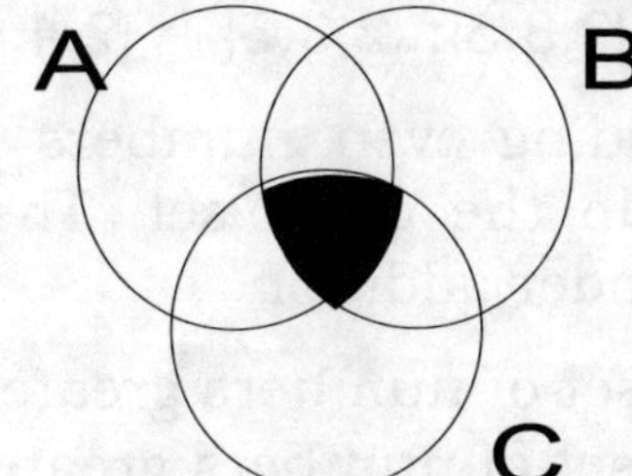

11. Based on the Venn diagram below, which of the following statements is true?

(A) $A \cup B$ = {2,4}
(B) A = {1,2,3,4,6,8}
(C) $A \cap B$ = {2,4}
(D) B = {1,2,3,4,6,8}

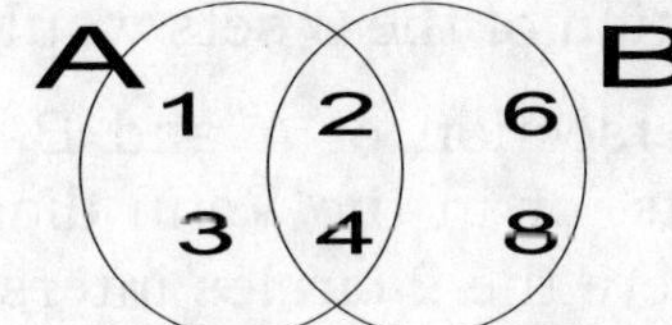

ANSWERS – SET THEORY

1. D — $A \cup B$ contains all of the elements included in either A or B.
2. D — $A \cap B$ contains only those elements common to both sets.
3. B — The intersection of B and C is {2,8}. The union of {2,8} with A = {1,2,3,4,5} is {1,2,3,4,5,8}.
4. C — There are 5 elements so there are 2^5 or 32 subsets including the null set and the set itself.
5. D — The first element must be from set B and the second element must be from set C. In choice D, both elements are from B.
6. C — Each of the 5 elements in set A can be paired with each of the 5 elements in set B. A X B has 5 x 5 or 25 ordered pairs.
7. A — $A \cap B$ is {2,3,5}; $\{2,3,5\} \cap \{2,4,6,7,8\}$ is {2}.
8. C — Since adding even numbers produced an even sum, all answers are within the given set. The set of even numbers is, therefore, closed under addition.
9. B — A is thc set of numbers greater than -1 and less than or equal to 3. B is the set of numbers greater than or equal to 3. $A \cup B$ is the set of numbers greater than -1.
10. B — The shaded part represents the elements common to A and B and C or $A \cap B \cap C$. Using the elements defined in problems 1 - 7, the intersection of the 3 sets would be {2}.
11. C — The intersection of A and B includes those elements common to both sets, or in the Venn diagram those elements included in the area where the 2 circles intersect. $A \cap B = \{2,4\}$.

LOGIC

An area of mathematics which utilizes many of the principles in set theory is logic. As in sets, the key to these problems is learning the symbols which are used. There are no real computations. Rather, logic tests your reasoning powers in problems in which English sentences have been replaced with symbols.

The basic unit of logic is the statement—a simple sentence, such as, "I am tall." In logic we are concerned with the compounding of simple statements. To do this we replace statements with letters.

♦ Let p represent the statement, "I am tall," and q represent the statement, "I am a good tennis player." The compound sentence, "I am tall, and I am a good tennis player," could be written as p and q. To simplify even further, we replace "and" by the symbol $\wedge$. Now we can write $p \wedge q$. This is known as the conjunction.

To write, "I am tall or I am a good tennis player," we write $p \vee q$. This is known as the **disjunction** of p and q. To write, "I am not tall," we use ~p. This is known as the **negation** of p. The negation of $p \wedge q$ would be $\sim(p \wedge q)$. This would translate as, "I am not both tall and a good tennis player." How would $\sim p \vee q$ translate? "I am not tall, or I am a good tennis player."

A truth table shows the truth or falsity of a statement using logic symbols:

EXAMPLE

p	~p	This table says:
T	F	If p is true, ~p is false.
F	T	If p is false, ~p is true.

The truth table for a compound sentence depends upon the truth of its components. The truth table for $p \wedge q$ would be:

p	q	$p \wedge q$	
T	T	T	If both p and q are true, $p \wedge q$ is true.
T	F	F	If p is true but q is false, $p \wedge q$ is false.
F	T	F	If p is false and q is true, $p \wedge q$ is false.
F	F	F	If both p and q are false, $p \wedge q$ is false.

The truth table for $p \vee q$ would be:

p	q	$p \vee q$	
T	T	T	If both p and q are true, $p \vee q$ is true.
T	F	T	If either p or q is true, $p \vee q$ is true.
F	T	T	If either p or q is true, $p \vee q$ is true.
F	F	F	If both p and q are false, $p \vee q$ is false.

Now let's take a look at two statements:

A: You have a press pass.

B: You can join the team on the field.

Combining statements A and B, we make the following sentence:

If you have a press pass, then you can join the team on the field.

This is called a conditional statement. It has a hypothesis or antecedent, "If you have a press pass" and a conclusion or consequent, "then you can join the team on the field". The conclusion that you join the team on the field is conditional. It won't always happen. Not everyone gets to go onto the field. And, I don't know if you have a pass. But let's just "hypothesize" that you do have one. Then logically I would conclude that you can join the team on the field.

Symbolically, we represent this conditional relationship as A → B. This can be read as "A implies B" or "If A, then B".

In conditional statements we also talk about necessary and sufficient conditions. In our example, having a press pass is sufficient for joining the team on the field. However, it's not necessary. You could be the coach or the ref, etc.

Suppose we reversed these two statements. The statement "If B, then A" or B → A is called the **converse** of "If A, then B". In our example, that would be "If you join the team on the field, then you have a press pass". As you can see, the converse is not necessarily true. Remember, it is sufficient but not necessary.

The statement "If not A, then not B" or ~A → ~B is called the **inverse** of "If A, then B". In our example, that would be "If you don't have a press pass, you can't join the team on the field". As with the converse, we know that this is not necessarily true.

Finally, the statement "If not B, then not A" or ~B → ~A is called the **contrapositive** of "If A, then B". In our example, that would be "If you can't join the team on the field, then you don't have a press pass". If the original statement is true, and in our example, it is, then the contrapositive is also true. If you are not on the field with the team, that tells us a lot. It tells us that you are not on the team, that you are not the coach, that you are not a ref, but it also tells us that you don't have a press pass. If you did have a pass, you would be there on the sideline, even if you were not the coach or a ref.

In summary:

The original conditional statement is written A → B.

The **converse** is written B → A.

The **inverse** is written ~A → not ~B.

The **contrapositive** is written ~B → ~A.

The original statement and its contrapositive are logically equivalent. If the original statement is true, then the contrapositive will also be true. If the original statement is false, then its contrapositive will also be false.

♦ The conditional statement "If you are not dressed, you can't go outside" is logically equivalent to which of the following?

(A) If you can't go outside, then you aren't dressed.
(B) If you are dressed, then you can go outside.
(C) If you aren't dressed, then you can go outside.
(D) If you can go outside, then you are dressed.

If we let A equal "You are dressed", and B equal "You can go outside", then the original statement can be represented as: $\sim A \rightarrow \sim B$. This is the contrapositive of and is logically equivalent to the statement $B \rightarrow A$. This is the statement "If you can go outside, then you are dressed". The correct answer is (D).

♦ A: You are a woman.

B: You can't become President of the USA.

The conditional statement $A \rightarrow B$ would be "If you are a woman, then you can't become President". This is false. If the hypothesis is true and the conclusion is false, then the conditional statement is false.

This is the only time that a conditional statement is logically false. If the hypothesis and conclusion are both true, then the conditional statement is true. Also, if the hypothesis is false, regardless of the conclusion, the conditional statement will be logically true.

p	q	$p \rightarrow q$	
T	T	T	If both antecedent and consequent are true, the conditional is true.
T	F	F	If the antecedent is true but the consequent is false, the conditional is false.
F	T	T	If the antecedent is false but the consequent is true, the conditional is true.
F	F	T	If both antecedent and consequent are false, the conditional is true.

Now, getting back to our original example:

A: You are a woman.

B: You can't become President of the USA.

The **contrapositive** is "If not B, then not A," or ~B → ~A. In our example, that would be "If you can become President, then you are not a woman". Again, the hypothesis is true but the conclusion is false, so the conditional statement is false.

The converse and the inverse have a similar logical relationship.

Now let's look at the statement:

> "You can go to the game on Sunday if and only if you help rake the leaves on Saturday."

Let A represent the statement "You can go to the game on Sunday"; and let B represent the statement "You help rake the leaves on Saturday". If we know that you can go to the game on Sunday, then we conclude that you must have raked the leaves on Saturday.

If A → B. Also, if we know that you raked leaves on Saturday, then we can conclude that you can go to the game on Sunday. If B → A. Raking the leaves is not only sufficient for going to the game, it is also necessary. No other good deed, no amount of money will make it possible for you to attend the game. So, if I hypothesize that you do, in fact, rake the leaves, then I can logically conclude that you can go to the game. Also, if I hypothesize that you can go to the game, then I can logically conclude that you must have raked the leaves.

An "if" statement implies a sufficient condition. An "only if" statement implies a necessary condition.

♦ Given the statement, "If Christmas is tomorrow, then it must be December", which of the following statements is the inverse?

(A) If tomorrow isn't Christmas, then it isn't December.
(B) If it is December, then tomorrow is Christmas.
(C) If it isn't December, then tomorrow isn't Christmas.
(D) If Christmas is tomorrow, then it isn't December.

A: Christmas is tomorrow.

B: It is December.

The inverse of "If A then B" is "If ~A, then ~B". Or in this case, If tomorrow isn't Christmas, then it isn't December. The correct answer is (A). (B) is If B, then A, which is the converse. (C) is If ~B, then ~A, which is the contrapositive.

PRACTICE PROBLEMS — LOGIC

1. If a represents "It is raining" and b represents "The streets are wet", which of the following represents the statement, "It is raining and the streets are wet"?

   (A) $a \vee b$
   (B) $a \wedge b$
   (C) $a \rightarrow b$
   (D) $a \rightarrow \sim b$

2. Which of the following is the converse of $a \rightarrow b$?

   (A) $a \rightarrow \sim b$
   (B) $b \rightarrow a$
   (C) $\sim a \rightarrow b$
   (D) $a \wedge b$

3. Which of the following is the negation of $p \vee q$?

   (A) $p \rightarrow q$
   (B) $q \rightarrow p$
   (C) $\sim p \wedge q$
   (D) $\sim(p \vee q)$

4. If $p \vee q$ is false, then which of the following must be false?

   (A) p
   (B) q
   (C) both p and q
   (D) neither p or q

5. Which of the following is the inverse of $a \rightarrow b$?

   (A) $\sim a \rightarrow \sim b$
   (B) $a \rightarrow \sim b$
   (C) $b \rightarrow a$
   (D) $\sim a \rightarrow b$

6. Which of the following is the contrapositive of $a \rightarrow b$?

   (A) $b \rightarrow a$
   (B) $\sim a \rightarrow b$
   (C) $\sim a \rightarrow \sim b$
   (D) $\sim b \rightarrow \sim a$

7. If p represents "I am a good student" and q represents "I will pass the test", which of the following represents "If I am not a good student, I will not pass the test"?

   (A) $p \rightarrow q$
   (B) $\sim p \rightarrow \sim q$
   (C) $\sim q \rightarrow \sim p$
   (D) $q \rightarrow p$

8. If the statement "All lawyers know Latin" is true, which of the following statements is implied from the first statement?

   (A) All non-lawyers do not know Latin.
   (B) If a person knows Latin, he is a lawyer.
   (C) If a person does not know Latin, he is a lawyer.
   (D) If a person does not know Latin, he is not a lawyer.

9. If p is true and q is false, then $p \vee q$ is:

   (A) false
   (B) true
   (C) equals $p \wedge q$
   (D) conditionally false

10. Given the conditional statement, $p \rightarrow q$, then $q \rightarrow p$ is

   (A) the converse
   (B) the inverse
   (C) the contrapositive
   (D) always true

ANSWERS – LOGIC

1. B The symbol $\wedge$ is called the conjunction and replaces the word "and".
2. B The converse switches the antecedent and the consequent.
3. D The symbol ~ is used for negation.
4. C Only when p and q are both false will the disjunction be false. In other words, if either one statement or the other is true, we consider the statement true.
5. A The inverse is the negation (~) of the conditional $a \rightarrow b$.
6. D The contrapositive is the negation of the converse.
7. B The statement represents the inverse.
8. D Only statement D is true if the original statement is true.
9. B If either p or q is true, then the disjunction is considered true.
10. A $q \rightarrow p$ is the converse of the conditional $p \rightarrow q$. The converse is not always true.

COORDINATE GEOMETRY

Coordinate Geometry deals with the location and relationship of points located on a two-dimensional surface. This surface is divided into 4 quadrants by two intersecting perpendicular lines called axes. The horizontal line is the x-axis and the vertical line is the y-axis. They are like horizontal and vertical number lines. Every point in this space is defined by an ordered pair of numbers, where the first number is always the distance traveled along the x-axis and the second number is always the distance traveled along the y-axis. The point where the x-axis and y-axis intersect is (0,0) and is called the origin.

Locate the points (5,2) and (-1,1).

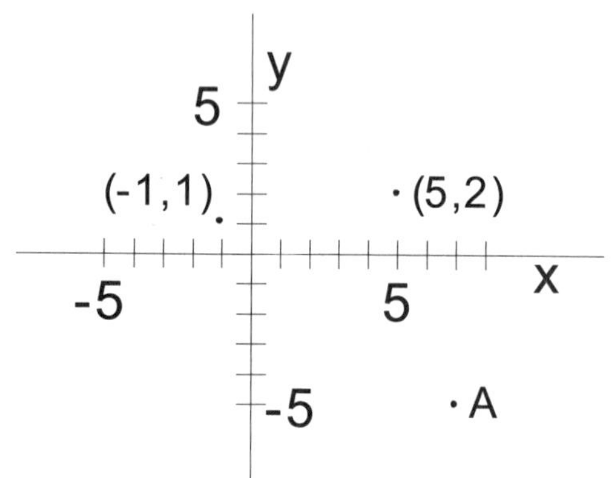

The point (5,2) is located by moving to the right 5 spaces from the origin and then up 2 spaces.

The point (-1,1) is located by moving to the left 1 space from the origin and then up 1 space.

EXAMPLE: What are the coordinates of point A?

(A) (7,5) (B) (7,-5)

(C) (-7, 5) (D) (-7,-5)

SOLUTION: To locate point A, you must move to the right 7 units from the origin and then down 5 units. That's a +7 and a -5. The point is (7, -5). The correct answer is (B).

Graphing a Straight Line

We can learn important information about a function by graphing it. The simplest way to graph a function is to find ordered pairs by substituting different values for the variables and finding the value of the function. We can then make a table and plot the points of the function.

To graph the algebraic equation y = 2x, we can set up a table of values by substituting values for x and solving for values of y:

x	y
0	0
1	2
2	4

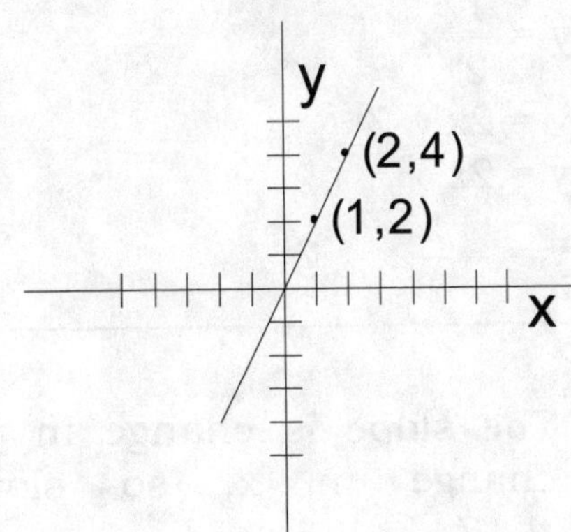

We can plot these points and connect them with a straight line as shown in the graph to the right of the table.

EXAMPLE: If f(x) = 2x + 4, then f(2) = 8, f(1) = 6, f(0) = 4, etc.

By plotting the points (2, 8), (1, 6), (0, 4), we can get a picture of the graph. As you can see below, this function is a straight line:

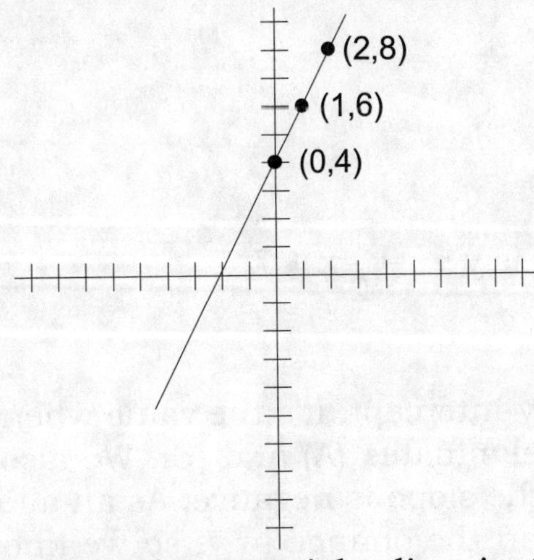

What we notice when we graph a straight line is that there is a constant relationship between the x and y coordinates, and in general, straight lines are represented by the equation $y = mx + b$ where m is the slope of the line, and b is the y-intercept. y is a linear function of x.

The y-intercept is the point where the line crosses the y-axis, or where x = 0.

In our equation, if we substitute 0 for x, we see that y = b.

The slope defines the slant of the line. In the equation y = 2x, we see that each time we move 1 in an x direction, we will move 2 in a y direction. m, written as a fraction, represents the change in y over the change in x; the larger the number, the steeper the slant. Positive slopes rise from left to right. Negative slopes fall from left to right.

♦ What is the equation of a line that passes through (0,7) and (6,4)?

(A) $y = \frac{1}{2}x$

(B) $y = 2x + 7$

(C) $y = 2x$

(D) $y = -\frac{1}{2}x + 7$

SOLUTION: The slope is change in y over change in x, so slope is $\frac{4-7}{6-0} = -\frac{1}{2}$. So, the equation is in the form $y = -\frac{1}{2}x + b$. Since either point must make this a true statement, we can substitute either in to solve for b.

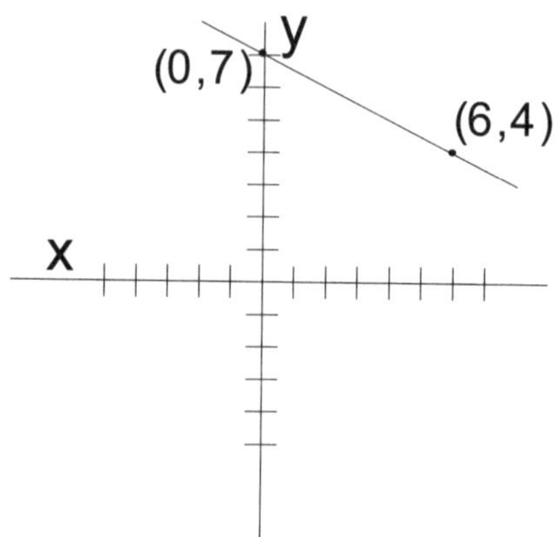

$$4 = -\frac{1}{2}(6) + b$$

$$4 = -3 + b$$

$$7 = b$$

The equation is $y = -\frac{1}{2}x + 7$.

The correct answer is (D).

The problem gave us the y-intercept; it's the value when x is 0, so we know that the value of b is 7. This eliminates (A) and (C). We also know that y decreases while x increase, so that the slope is negative. As an alternative, we see that the change in x is greater than the change in y, so we know that the slope is less than 1. Either way we eliminate (B). It has a positive slope greater than 1.

♦ If f(x) is a linear function such that f(-4) = 5 and f(2) = 8, then f(0) =

(A) 4
(B) 7
(C) 5
(D) 9

The slope of the line (m) is equal to the change in y over the change in x. This is equal to $\frac{5-8}{-4-2} = \frac{-3}{-6} = \frac{1}{2}$. So, the equation of our line is equal to $f(x) = \frac{1}{2}x + b$. We know that f(2) = 8, so b must equal 7. Our equation is $f(x) = \frac{1}{2}x + 7$. f(0) = 7.

The correct answer is (B).

Graphing Quadratic Functions

The graph of a function in which the variable has an exponent of 2 usually yields a curve, such as a circle, a parabola, an ellipse, or a hyperbola. By plotting enough pairs of points, you can get a rough "sketch" of the curve.

EXAMPLE: To graph $y = f(x) = x^2 + 2$:

x =	0	1	-1	2	-2
y =	2	3	3	6	6

As you can see, the same number for x positive or negative gives equal values of y. Plotting the points, we would get a rough sketch as shown below:

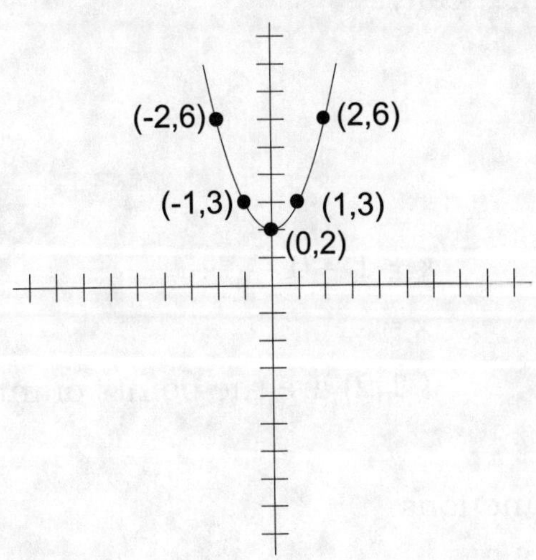

You can see in the function $f(x) = x^2 + 2$, the curve is always above the x-axis; therefore, it has no real number x-intercepts. The y-intercept is the point when x = 0. In $f(x) = x^2 + 2$, this point is (0, 2).

The point where two functions cross is an ordered pair (coordinates), which satisfy both functions. That is, substituting the first number for x and the second for y will make true statements for both functions. To find their intersection points, you can set the two functions equal to each other and solve. Because this example involves a quadratic function, you will need to factor the equation and solve for the two roots.

♦ If the two functions $y = f(x) = x^2 + 1$ and the function $y = f(x) = x + 3$ are both graphed, they will cross at the points (2, 5) and (-1, 2). Here is the graph of $y = f(x) = x^2 + 1$ and $y = f(x) = x + 3$:

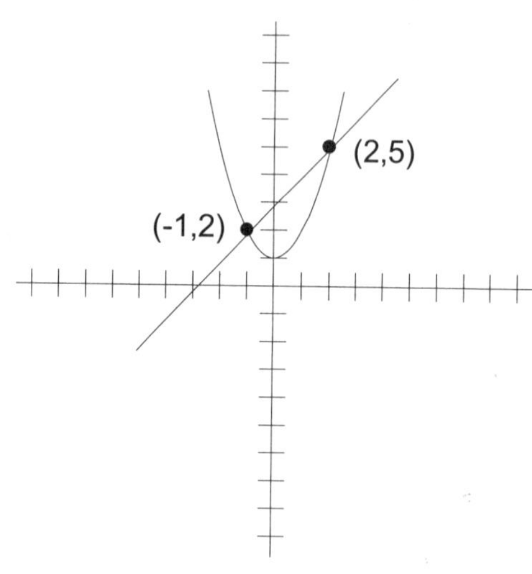

Finding the roots mathematically:

$x^2 + 1 = x + 3$

$x^2 - x - 2 = 0$

$(x - 2)(x + 1) = 0$

$x - 2 = 0$ AND $x + 1 = 0$

$x = 2$ $x = -1$

(2,5) AND (-1,2) are the points of intersection

Both pairs satisfy both functions:

for:	$y = x^2 + 1$			$y = x + 3$	
(2, 5)	$5 = 2^2 + 1$	True	(2, 5)	$5 = 2 + 3$	True
(-1, 2)	$2 = (-1)^2 + 1$	True	(-1, 2)	$2 = -1 + 3$	True

PRACTICE PROBLEMS — COORDINATE GEOMETRY

1. Which of the following numbered graphs could represent the graph of $y = 2x+4$?

(A) 1
(B) 2
(C) 3
(D) 4

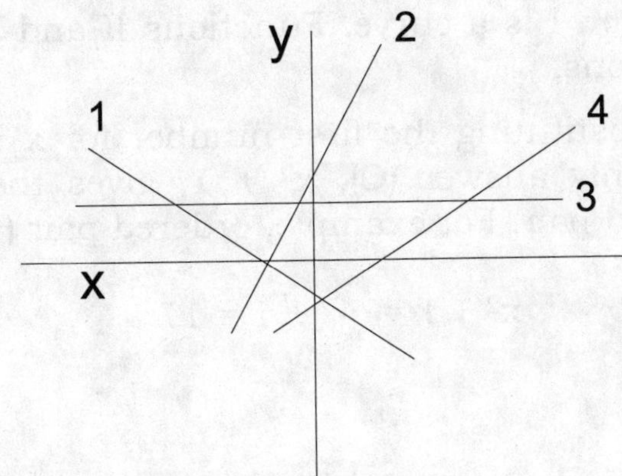

2. Which of the following functions would graph as a curve?

I. $y = x^2 + 2$

II. $y = 2x$

III. $y = 3(x - 2)$

(A) I only
(B) I and II
(C) I and III
(D) III only

3. Which of the following functions fits the ordered pairs shown?

$f(x, y) = \{(0, 1), (2, 5), (4, 17)\}$

(A) $f(x) = 2x$
(B) $y = x + 3$
(C) $f(x) = x^2 + 1$
(D) $f(x) = x^2$

ANSWERS – COORDINATE GEOMETRY

1. B — For $y = 2x + 4$, the slope is 2 and the y-intercept is 4. Only line 2 has a positive slope and a positive y-intercept. Equations 1 and 4 have negative y-intercepts and equation 3 has a slope of zero.

2. A — Function I is a curve. Functions II and III are linear (straight-line) equations.

3. C — By substituting the first number for x in each function, you find that only answer (C), $x^2 + 1$, gives the second element of each ordered pair. For example, ordered pair (4,17): $x = 4$, $y = 17$

$$x^2 + 1 = 4^2 + 1 = 17$$

COMBINATIONS, PERMUTATIONS & PROBABILITY

Permutations

The **permutations** of a set of objects are the number of different subsets you can form from a set, where order is important.

♦ How many different ways can a set of 6 different colored flags be arranged?

There are six different ways to fill the first position. Once the first position is filled, there are 5 flags left, so there are 5 choices for the second position, 4 for the third, and so on.

6* 5 * 4 * 3 * 2 * 1

Each way the first position can be filled is paired with each way the second position can be filled, etc. This pairing implies multiplication, and, in fact, the total number of permutations is

6 * 5 * 4 * 3 * 2 * 1 = 720 permutations.

This product can also be written as **6!** (mathematically termed 6 factorial).

♦ From the same set of 6 different colored flags, how many different arrangements of three flags can be made?

6 * 5 * 4 = 120

This could also be written as $\frac{6\cdot5\cdot4\cdot3\cdot2\cdot1}{3\cdot2\cdot1} = \frac{6!}{3!}$

In general, the formula for the permutations of n things taken r at a time, or ${}_nP_r$, is $\frac{n!}{(n-r)!}$, where n equals the total number of objects and r equals the number chosen.

♦ Marisa bought 5 new blouses, 3 new skirts, and 2 new pairs of shoes for her new job. If she works Monday through Friday, how many weeks can she go without wearing the same outfit twice?

(A) 1
(B) 3
(C) 6
(D) 2

First, we have to find how many different outfits she can make. She has 5 choices for the blouse, which can be paired with any one of 3 choices for the skirt, which, in turn, can be paired with either of the 2 choices for the shoes.

$5 * 3 * 2 = 30$

She can make 30 different outfits. But how many weeks can she go without repeating an outfit? That would be $\frac{30}{5}$ or 6 weeks; (C).

Combinations

Combinations are the number of different subsets you can form from a set of objects, where order is not important.

♦ The Senior Class Council is made up of one representative from each of the 12 senior homerooms. They are going to pick 3 members to serve as a subcommittee to plan the Senior Prom. How many different subcommittees are possible?

(A) $\frac{12!}{9!3!}$

(B) $\frac{12!}{9!}$

(C) $\frac{12!}{3!}$

(D) $\frac{12!}{9!(9-3)!}$

If order were important in the selection, we would have 12 * 11 * 10. But that would be counting every three-member panel 3! times. The committee of Fred, Sally, and George and the committee of Sally, Fred, and George would be considered two different committees if order were important. However, we know that this is the same committee, regardless of the order in which we picked the 3 names. So, we must divide 12 * 11 * 10 by 3! to eliminate the repetition.

The number of subcommittees $= \frac{12 \cdot 11 \cdot 10}{3!} = \frac{12 \cdot 11 \cdot 10 \cdot 9!}{3!9!} = \frac{12!}{3!9!}$.

The correct answer is (A). When multiplied out, it is 220.

In general, the formula for the combination of n things taken r at a time, or ${}_nC_r$, is:

$${}_nC_r = \frac{n!}{r!(n-r)!}$$

♦ How many different hands of 5 cards can you deal from a 52-card deck?

The order that you receive the cards is unimportant. The only thing you care about what 5 cards you have at the end of the deal.

We already know that if order were important, we have ${}_{52}P_5 = \frac{52!}{47!}$. What is the impact of considering order? You are counting each hand of 5 cards 5! times. If you divide the number of permutations by 5!, you will have the number of combinations.

$$ {}_{52}C_5 = \frac{52!}{47!*5!} = \frac{52*51*50*49*48*47!}{47!*5!} = \frac{52*51*50*49*48}{5*4*3*2*1} $$
$$ = 13*17*10*49*24 = 2{,}598{,}960 $$

Probability

The probability that an event will occur is the ratio of the number of favorable outcomes to total number of outcomes.

♦ If you roll a die, what is the probability that you will roll an even number?

The set of favorable outcomes = {2, 4, 6}
The set of total outcomes = {1, 2, 3, 4, 5, 6}

$$ P(\text{even}) = \frac{3}{6} = \frac{1}{2} $$

♦ What is the probability that you draw an ace from a 52-card deck?

There are four favorable outcomes = the ace of spades, hearts, diamonds, & clubs. There are 52 total outcomes.

$$ P(\text{ace}) = \frac{4}{52} = \frac{1}{13} $$

The probability of not getting an ace is $\frac{48}{52}$ or $\frac{12}{13}$. The probability that an event doesn't occur is equal to 1 - P(event).

♦ What is the probability that a single draw from a standard 52-card deck will yield a number divisible by 3?

(A) $\frac{3}{52}$

(B) $\frac{1}{13}$

(C) $\frac{3}{13}$

(D) $\frac{2}{13}$

There are 3 numbers divisible by 3 (3, 6, and 9) and there are 4 of each in the deck, so there are 12 favorable outcomes. There are 52 total outcomes. The probability is $\frac{12}{52}$ or $\frac{3}{13}$. The correct answer is (C).

Probability of a Combination of Events

We've discussed the probability of a single event occurring. What about the probability that one of two independent events, either event A or event B, will occur? The word "or" in probability translates to addition.

♦ What is the probability that you draw an ace or a king from a 52-card deck?

This is equal to the probability that you draw an ace plus the probability that you draw a king.

$$P(\text{ace or king}) = P(\text{ace}) + P(\text{king}) = \frac{1}{13} + \frac{1}{13} = \frac{2}{13}.$$

The key here is that a card couldn't be both an ace and a king.

In a situation such as the probability that the card is an ace or a heart, you would be better off counting the favorable outcomes. There are 4 aces, one of which is a heart, and there are 12 other hearts. So, there are a total of 16 favorable outcomes out of a total of 52 outcomes. The probability that a card is either an ace or a heart, P(ace or heart), is equal to $\frac{16}{52}$, or $\frac{4}{13}$

The word "and" in probability translates to multiplication.

♦ If you deal 2 cards from a 52 card deck, what is the probability that both are aces?

This is equal to the probability that you draw an ace on the first card times the probability that you draw an ace on the second card (given that you already drew an ace on the first card).

P(2 aces) = P(ace on first card) * P(ace on second card) =

$$\frac{4}{52} * \frac{3}{51} = \frac{12}{2652}$$

The key here is that the first card is not replaced in the deck, so there are only 3 aces remaining for the second card and only 51 cards remaining in the deck.

♦ The probability of dealing an ace or a heart from a standard 52 card deck is:

(A) $\frac{17}{52}$

(B) $\frac{4}{13}$

(C) $\frac{1}{4}$

(D) $\frac{1}{52}$

There are 13 hearts, one of which is the ace of hearts. There are 3 other aces that are not hearts. So there are a total of 16 favorable outcomes out of a total of 52. So, the probability of dealing an ace or a heart is $\frac{16}{52} = \frac{4}{13}$. The correct answer is (B).

$\frac{17}{52}$ counts the ace of hearts twice; $\frac{1}{4}$ or $\frac{13}{52}$ only counts the favorable outcomes that are hearts; $\frac{1}{52}$ is the probability that the card is an ace **AND** a heart.

♦ A fair die is rolled and a card is drawn from a standard deck of 52 cards. What is the probability that the top face of the die will show an even number and the card will be a club?

(A) $\frac{1}{8}$

(B) $\frac{3}{4}$

(C) $\frac{1}{2}$

(D) $\frac{1}{4}$

This is the probability of a combination of events, so we will multiply the individual probabilities together. $P(\text{even}) = \frac{3}{6} = \frac{1}{2}$. $P(\text{club}) = \frac{13}{52} = \frac{1}{4}$. $P(\text{even and club}) = \frac{1}{2} * \frac{1}{4} = \frac{1}{8}$. The correct answer is (A).

Conditional Probability

Conditional probability is the probability of one outcome given or based on successfully completing or satisfying another condition or event.

♦ In a recent exit poll, 6 out of 10 women said that education was a more important issue than taxes, but only 3 out of 10 men agreed. What is the probability that the person surveyed thought education was the more important issue, given that the person was a woman?

Below is a table depicting the M (men) and W (women), and their choice of issue E (education) and T (taxes) with row and column totals:

	E	T	
M	3	7	10
W	6	4	10
	9	11	20

Given that the person is a woman, there are 10 total outcomes, of which 6 would be considered successful. The probability is $\frac{6}{10}$ or $\frac{3}{5}$.

♦ Based on the above, given that a person believes the more important issue is taxes, what is the probability that the person is a man?

(A) $\frac{7}{11}$

(B) $\frac{7}{10}$

(C) $\frac{1}{3}$

(D) $\frac{3}{10}$

There are 11 favorable outcomes (people believing that the more important issue is taxes. Of these, 7 are men. The probability is $\frac{7}{11}$. This can be written as $P(M|T) = \frac{7}{11}$. The correct answer is (A).

$\frac{7}{10}$ is the probability that the person believes the more important issue is taxes, given the person is a man.

$\frac{1}{3}$ is the probability that the person is a man, given that the person believes the more important issue is education.

$\frac{3}{10}$ is the probability that the person believes education is the more important issue, given the person is a man.

In general, we write conditional probability, the probability of X given Y as $P(X|Y) = n(X \cap Y)/n(Y)$. In our example, the number of people who are men and believe taxes are the primary issue is 7. The number of people who think taxes are the primary issue is 11.

PRACTICE PROBLEMS – PROBABILITY, COMBINATIONS & PERMUTATIONS

1. What is the probability of drawing a picture card from a 52-card deck?

   (A) $\frac{1}{13}$

   (B) $\frac{4}{13}$

   (C) $\frac{1}{52}$

   (D) $\frac{3}{13}$

2. What is the probability that you can flip a coin three times and get heads each time?

   (A) $\frac{1}{8}$

   (B) $\frac{1}{2}$

   (C) $\frac{1}{4}$

   (D) $\frac{3}{8}$

3. In a game of 21 (blackjack) you have a 10 and a 3. What is the probability that your next card will give you a total of 19, 20, or 21 (assume the 10 and 3 are the only cards drawn from the deck so far)?

   (A) $\frac{3}{13}$

   (B) $\frac{6}{25}$

   (C) $\frac{3}{52}$

   (D) $\frac{3}{50}$

4. What is the probability that it will not rain on a given day if the probability that it will rain is 1/3?

(A) $\frac{1}{3}$

(B) $\frac{1}{6}$

(C) $\frac{2}{3}$

(D) $\frac{1}{2}$

5. If you draw one of 24 slips of paper numbered consecutively from 1 - 24, what is the probability of drawing a number exactly divisible by 3?

(A) $\frac{2}{3}$

(B) $\frac{1}{24}$

(C) $\frac{5}{8}$

(D) $\frac{1}{3}$

6. What is the probability that a random draw of a card from a deck is a 9 or 8?

(A) $\frac{8}{13}$

(B) $\frac{2}{13}$

(C) $\frac{1}{13}$

(D) $\frac{4}{13}$

7. If you are to choose a committee of 4 people from a group of 8, how many different committees are possible?

(A) 100
(B) 32
(C) 64
(D) 70

8. How many different telephone area codes can be formed using the digits 0 - 9, if 0 isn't used for the first digit?

(A) 500
(B) 1000
(C) 900
(D) 729

9. How many triangles can be formed from a set of 8 different points, no three of which lie in a straight line?

(A) 336
(B) 60
(C) 70
(D) 56

10. How many even three-digit numbers exist?

(A) 500
(B) 400
(C) 450
(D) 100

ANSWERS – PROBABILITY, COMBINATIONS & PERMUTATIONS

1. D There are 12 picture cards in a deck. Therefore, there are $\frac{12}{52}$ or $\frac{3}{13}$ chances of getting a picture card.

2. A The probability of getting heads 3 times would be $\frac{1}{2} \times \frac{1}{2} \times \frac{1}{2}$ or $\frac{1}{8}$.

3. B If the next card drawn were a 6, 7, or 8, you would have a total of 19, 20, or 21. Since there are 4 of each, there are a total of 12 favorable cards. Since there are 50 cards left, the probability would be $\frac{12}{50}$ or $\frac{6}{25}$.

4. C. The probability that an event will happen, plus the probability it will not, must give a sum of 1. Therefore, the probability it will not happen is $1 - \frac{1}{3}$ or $\frac{2}{3}$.

5. D There are 8 numbers from 1-24 which are divisible by 3. The probability is, therefore, $\frac{8}{24}$ or $\frac{1}{3}$.

6. B The probability of getting a 9 or 8 is $\frac{8}{52}$, or $\frac{2}{13}$.

7. D The number of committees would be ${}_8C_4 = \frac{8 \times 7 \times 6 \times 5}{4 \times 3 \times 2 \times 1}$ or 70.

8. C For the first digit, you have a choice of 9 digits (1-9). For the second and third digits you have a choice of 10 digits each. Therefore, there are 9 × 10 × 10 or 900.

9. D Each triangle requires 3 points. From 8 points there are ${}_8C_3$ possible triangles. ${}_8C_3 = \frac{8 \times 7 \times 6}{3 \times 2 \times 1} = 56$.

10. C The first digit can be 1-9 (nine possible digits); the second, 0-9 (ten possible digits); the third can only be 0, 2, 4, 6, or 8 (5 possible digits). Then there are 9 × 10 × 5 or 450 even 3-digit numbers.

NUMBER BASES

Our number system is called base 10 because it is composed of ten digits; the numbers 0 to 9. A multi-digit number represents powers of 10.

The number 584 represents $(5 \times 10^2) + (8 \times 10^1) + (4 \times 10^0)$.

Other number bases are often used in mathematics. The computer is really a base 2 machine. The base 2 is composed of the numbers 0 and 1. A base 5 number system uses the numbers 0, 1, 2, 3, 4. Some of the CLEP examination problems might ask you to represent a base 5 number, for example, in base ten.

♦ The number 234_5 (read two-three-four base five) represents:

$$= (2 \cdot 5^2) + (3 \cdot 5^1) + (4 \cdot 5^0)$$
$$= 2(25) + 3(5) + 4(1)$$
$$= 50 + 15 + 4$$
$$= 69 \text{ in base } 10$$

The number 10101_2 could be changed to its base 10 equivalent in the same way.

♦ What is the base 10 equivalent of 10101_2?

$$= (1 \cdot 2^4) + (0 \cdot 2^3) + (1 \cdot 2^2) + (0 \cdot 2^1) + (1 \cdot 2^0)$$
$$= 16 + 0 + 4 + 0 + 1$$
$$= 21$$

To change a base 10 number into another base equivalent, divide by powers of the base in descending order.

♦ Change 134 into its base 6 equivalent.

Divide by the largest power of 6 possible:

$6^2 = 36$ then: $36\overline{)134}$ = 3, 108, 26. There are 3 powers of 6^2.

And: $6^1 = 6$ Then: $6\overline{)26}$ = 4, 24, 2. There are 4 powers of 6^1.

There are 2 powers of 6^0.

Therefore, 134_{10} is the equivalent of 342_6. To prove, change back to base 10. Thus 342_6 is equivalent to:

$3(6^2) + 4(6^1) + 2(6^0) = 3(36) + 4(6) + 2(1) = 108 + 24 + 2 = 134.$

LOGARITHIMS

A **log** is an exponent of a number. For example, to say $2^3 = 8$, we can write $\log_2 8 = 3$, which is read "the log of eight base two is three." Logs can be of any number base. $\text{Log}_a x = b$ is the same thing as saying, in exponential form, $a^b = x$. The common logarithms are to the base 10. Thus, since $10^2 = 100$, $\log_{10} 100 = 2$. If the base is not written, it is understood that the base is 10. So, log 14 = x says $10^x = 14$. You can see that any number between powers of 10 will have a log that is part decimal.

$10^1 = 10$, therefore log 10 = 1

$10^2 = 100$, therefore log 100 = 2

$10^3 = 1000$, therefore log 1000 = 3

$10^? = 25$, therefore log 25 = a number between 1 and 2

Logarithms of numbers can be found in tables in most mathematics books that deal with the topic. For purposes of the exam you need only approximate. Most of the examination questions deal with the properties of logs in general rather than specific numbers. Logs are used to facilitate computation with large and small numbers. The properties of logs that make them a useful tool are as follows:

$$\textbf{Log AB = Log A + Log B}$$

$$\textbf{Log A} \div \textbf{B = Log A - Log B}$$

$$\textbf{Log A}^{\textbf{b}} \textbf{ = b Log A}$$

EXAMPLE: To compute $\frac{5{,}234 \times (3{,}758)^2}{2{,}714}$, we could find the log of 5,234, add to it twice the log of 3,758, and subtract the log of 2,714. This gives us the log of the answer, called the anti-log. Using the log table we can find the number which has that log. In doing the log problems on the examination it is important to remember that logarithms are really exponents.

PRACTICE PROBLEMS — LOGARITHMS

1. Represent the following in exponential form:

   A. $\log_2 4 = 2$

   B. $\log_3 81 = 4$

   C. $2 = \log_4 16$

   D. $\log_9 3 = \frac{1}{2}$

   E. $\log_b A = x$

2. Represent the following in log form:

   A. $2^3 = 8$

   B. $3^3 = 27$

   C. $4^{1/2} = 2$

   D. $A^x = B$

   E. $8^{1/3} = 2$

3. Find the value of x:

   A. $\log_x 216 = 3$

   B. $\log_x 81 = 4$

4. Show the following are true:

   A. $\log_2 64 = 3 \log_8 64$

   B. $\log_2 8 \cdot \log_8 2 = 1$

   C. $\log_3 27 \cdot \log_{27} 3 = 1$

5. If log 758 = 2.8797 and log .416 = (9.6191 - 10), find log (758 · .416).

6. If log 2 = .3010, find log of 2^5.

ANSWERS - LOGARITHIMS

1. Log Form: Exponential Form:

	Log Form:	Exponential Form:
A.	$\log_2 4 = 2$	$2^2 = 4$
B.	$\log_3 81 = 4$	$3^4 = 81$
C.	$2 = \log_4 16$	$4^2 = 16$
D.	$\log_9 3 = \frac{1}{2}$	$9^{\frac{1}{2}} = 3$ (Remember $9^{\frac{1}{2}} = \sqrt{9} = 3$)
E.	$\log_b A = x$	$b^x = A$

2. Exponential Form: Log Form:

	Exponential Form:	Log Form:
A.	$2^3 = 8$	$\text{Log}_2 8 = 3$
B.	$3^3 = 27$	$\text{Log}_3 27 = 3$
C.	$4^{1/2} = 2$	$\text{Log}_4 2 = \frac{1}{2}$
D.	$A^X = B$	$\text{Log}_A B = X$
E.	$8^{1/3} = 2$	$\text{Log}_8 2 = \frac{1}{3}$

3. A. $\log_x 216 = 3$, then $x^3 = 216$ and $x = \sqrt[3]{216} = 6$ (because $6 \cdot 6 \cdot 6 = 216$)

   B. $\log_x 81 = 4$, then $x^4 = 81$ and $x = \sqrt[4]{81} = 3$ (because $3 \cdot 3 \cdot 3 \cdot 3 = 81$)

4. Show the following are true:

   A. $\text{Log}_2 64 = x$ $\qquad$ $3\ \text{Log}_8 64 = x$

   $2^x = 64$ $\qquad$ $8^x = 64$

   $x = 6$ $\qquad$ $x = 2$

   $(2 \cdot 2 \cdot 2 \cdot 2 \cdot 2 \cdot 2 = 64)$ $\qquad$ $3(2) = 6$

B. $\text{Log}_2 8 \bullet \text{Log}_8 2 = 1$

$2^x = 8 \qquad 8^x = 2$

$x = 3 \qquad x = \frac{1}{3}$

$3 \bullet \frac{1}{3} = 1$

C. $\text{Log}_3 27 \bullet \text{Log}_{27} 3 = 1$

$3^x = 27 \qquad 27^x = 3$

$x = 3 \qquad x = \frac{1}{3}$

$3 \bullet \frac{1}{3} = 1$

This example is true for all problems where the base and number are switched. In general, $\text{Log}_b A \cdot \text{Log}_a B$ will always equal 1

(i.e., $\text{Log}_3 12 \cdot \text{Log}_{12} 3 = 1$).

5. In this example the value of the logs was obtained from a log table.

Using the Law of Logs which says Log AB = Log A + Log B:

$$\begin{aligned}\text{Log}\,(758 \cdot .416) &= 2.8797 + (9.6191 - 10)\\ &= 2.8797\\ &\quad \underline{(9.6191 - 10)}\\ &\quad 12.4988 - 10\\ &= 2.4988\end{aligned}$$

If you look up the number .4988 in a log table, you will find it represents the number 3153. With a characteristic of two, this would represent the number 315.3. Multiply the numbers 758 × .416 to prove this is correct (approximately).

6. Using the Law of Logs:

$$\begin{aligned}\text{Log}\, a^b &= b \log a\\ \text{Log}\, 2^5 &= 5 \log 2\\ &= 5(.3010)\\ &= 1.5050\end{aligned}$$

If you look up .5050 in a log table, you will find that it represents the number 32. Does $2^5 = 32$?

IMAGINARY AND COMPLEX NUMBERS

Since there is no real number to express the square root of a negative number, mathematicians use the letter "i" to represent $\sqrt{-1}$. The number i is called the imaginary unit. By using i we can express $\sqrt{-8}$, for example, as $i\sqrt{8}$. Thus $\sqrt{-16}$ could be represented by 4i. By using i we can perform operations with real and imaginary numbers. To facilitate operations with imaginary numbers it is helpful to see the effect of exponents on i.

$$i^1 = i = \text{definition}$$

$$i^2 = -1 = \text{definition}$$

$$i^3 = i^2 \cdot i = -1(i) = -i$$

$$i^4 = i^2 \cdot i^2 = -1 \cdot -1 = 1$$

Each succeeding power will repeat the pattern i, -1, -i and 1.

The examination may require you to simplify expressions involving imaginary units.

EXAMPLE: $(3 + \sqrt{-5})(3 - \sqrt{-5}) = 3^2 - (\sqrt{-5})^2$

$$= 3^2 - (i\sqrt{5})^2 = 3^2 - i^2(\sqrt{5})^2$$

$$= 9 - 5i^2$$

$$= 9 - 5(-1)$$

$$= 9 + 5 = 14$$

Numbers composed of a real number and an imaginary part are known as complex numbers. The general form of a complex number is a + bi, where a is the real part and b the imaginary part of a + bi, a and b being real numbers. To perform operations with complex numbers think of i as a variable and change any expression with powers of i by using the relationships shown above.

♦ $(3 + 2i)(7 + 4i) = (3)(7) + (3)(4i) + (7)(2i) + (2i)(4i)$

$$= 21 + 12i + 14i + 8i^2$$

$$= 21 + 26i + 8(-1)$$

$$= 21 + 26i - 8$$

$$= 13 + 26i$$

PRACTICE PROBLEMS — IMAGINARY AND COMPLEX NUMBERS

1. Multiply:

   A. $2(5i)$

   B. $(7i)(6i)$

   C. $(-15i)(\frac{1}{3}i)$

   D. $\sqrt{-9} \cdot \sqrt{-3}$

2. Add:

   A. $\sqrt{-16} + \sqrt{-49}$

   B. $5\sqrt{-48} - \frac{1}{3}\sqrt{-27}$

   C. $5\sqrt{-2} + \sqrt{-8}$

3. Simplify:

   A. i^9

   B i^{14}

   C. $\frac{8\sqrt{-32}}{2\sqrt{-2}}$

   D. $(2 + 3i)(3 - 2i)$

   E. $(3 - i)(4 + i)$

ANSWERS – IMAGINARY AND COMPLEX NUMBERS

1. Multiply:

   A. $2(5i) = 10i$

   B. $(7i)(6i) = 42i^2 = 42(-1) = -42$

   C. $(-15i)(\frac{1}{3}i) = -5i^2 = -5(-1) = 5$

   D. $\sqrt{-9} \cdot \sqrt{-3} = 3i \quad i\sqrt{3}$

   $= 3i^2\sqrt{3}$

   $= 3(-1)\sqrt{3}$

   $= -3\sqrt{3}$

2. Add:

A. $\sqrt{-16}+\sqrt{-49}$ = 4i + 7i = 11i

B. $5\sqrt{-48}-\frac{1}{3}\sqrt{-27} = 5i\sqrt{48}-\frac{1}{3}i\sqrt{27}$

$= 5i\sqrt{16}\cdot\sqrt{3}-\frac{1}{3}i\sqrt{9}\cdot\sqrt{3}$

$= 20i\sqrt{3}-i\sqrt{3}$

$= 19i\sqrt{3}$

C. $5\sqrt{-2}+\sqrt{-8} = 5i\sqrt{2}+i\sqrt{8}$

$= 5i\sqrt{2}+i\sqrt{4}\sqrt{2}$

$= 5i\sqrt{2}+2i\sqrt{2}$

$= 7i\sqrt{2}$

3. To simplify these examples use the following:

i^1 = i = definition

i^2 = -1 = definition

$i^3 = (i^2)\,i = -1(i) = -i$

$i^4 = (i^2)(i^2) = (-1)(-1) = 1$

A. $i^9 = (i^4)(i^4)(i) = (1)(1)(i) = i$

B $i^{14} = (i^4)(i^4)(i^4)(i^2) = (1)(1)(1)(-1) = -1$

C. $\frac{8\sqrt{-32}}{2\sqrt{-2}}=\frac{8i\sqrt{32}}{2i\sqrt{2}}=\frac{8i\sqrt{16}\sqrt{2}}{2i\sqrt{2}}=\frac{32i\sqrt{2}}{2i\sqrt{2}}=16$

D. $(2 + 3i)(3 - 2i) = (2)(3) + (2)(-2i) + (3i)(3) + (3i)(-2i)$

$= 6 + (-4i) + 9i + (-6i^2)$

$= 6 - 4i + 9i + (-6)(-1)$

$= 6 + 5i + 6 = 12 + 5i$

E. $(3 - i)(4 + i) = (3)(4) + (3)(i) + (-i)(4) + (-i)(i)$

$= 12 + 3i - 4i - i^2$

$= 12 - i - (-1)$

$= 13 - i$

PREPARING TO TAKE THE CLEP MATHEMATICS EXAMINATION

Now that you have finished the entire review section, you are ready to take a practice test. This practice test follows the format of the CLEP General Examination in Mathematics. We have included a preliminary section of warm-up exercises for you to do first. If you have difficulty with these problems, go back and review the information in the book before taking the practice examination.

It is important that you take this test when you have completed the warm-up exercises comfortably. While you may have done well on each type of question in isolation in the preceding section of the book, another skill is necessary. You must practice working with a variety of questions. In effect, you need to practice "switching gears" or training your mind to go from one type of question to another in a short period of time.

For best results, try to simulate the test situation as nearly as possible. The following procedure should be followed for maximum benefit:

1. Find a quiet spot where you won't be disturbed.
2. Time yourself accurately. Work 45 minutes in Part A and 45 minutes in Part B. Don't quit until the time is up!
3. Use the coding system for a systematic approach to the examination.

After you finish:

1. Check your answers.
2. Review procedures for doing questions if a particular type is giving you trouble.

WARM-UP EXERCISE

Before you get started with the actual practice exam in the next section, do the 15 problems in the warm-up exercise. This is not a timed exercise. However, if you struggle through these problems, you should consider additional review before attempting the practice examination.

1. A person can choose any 2 of 8 items from menu list I and any 3 of 7 items from menu list II. Approximately how many different meal combinations are possible?
   (A) 1,000
   (B) 500
   (C) 1,500
   (D) 750

2. In the conditional p --> q, p is true and q is false. What can be said about $q \rightarrow p$?
   (A) $q \rightarrow p$ is false
   (B) $q \rightarrow p$ is true
   (C) $p \rightarrow q$ is true
   (D) $p \wedge q$ is true

3. If for real numbers (x,y), $f(x) = x^2 + 3x - 1$, $f(x - 2) =$

   (A) $x^2 + 3x - 2$
   (B) $x^2 - x + 3$
   (C) $x^2 - x - 3$
   (D) $2x^2 - 3x + 3$

4. Which of the following operations are commutative?
   I. Addition
   II. Subtraction
   III. Multiplication
   IV. Division

   (A) I and II
   (B) I and III
   (C) II and IV
   (D) I, II, III, and IV

5. If $\sqrt{-1}$ is represented by i, what is $\sqrt{-8} \times \sqrt{-2}$?

   (A) 4i
   (B) 2i
   (C) -4
   (D) 16i

6. The equation of the graph shown below could only represent which of the following equations?

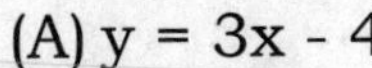

(A) $y = 3x - 4$
(B) $y = x^2 - x$
(C) $y = x^3 - x$
(D) $3y = x - 4$

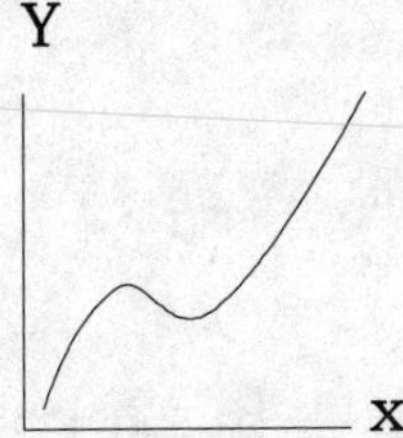

7. If A = {b, c, d} and B = {6, 7}, how many elements are in A x B?
(A) 3
(B) 6
(C) 5
(D) 2

8. Change 132_5 to base 10 notation:
(A) 52
(B) 48
(C) 65
(D) 42

9. A person starts a chain letter by writing two friends and requesting that each send a copy to two others. If the chain is unbroken after the fourth set is mailed, what is the total spent for postage at $.32 a letter?

(A) $9.92
(B) $9.60
(C) $5.12
(D) $8.96

10. Five students receive test scores of 62, 84, 99, 77, and 88. What is the average of the mean and the median scores?

(A) 82
(B) 83
(C) 85
(D) 86

11. Which of the following is equivalent to $\sqrt{\sqrt{a^{12}}}$?

(A) a^3
(B) a^4
(C) a^2
(D) a^6

12. A bicycle lock has a combination consisting of three numbers. If only the numbers 0 through 5 are on the dial, how many possible combinations are there?

(A) 216
(B) 15
(C) 125
(D) 16

13. For the function $y = 2x + 1$, $x = \{0, 1, 2\}$, what is the domain of the inverse of the function?

(A) $\{1, 3, 5\}$
(B) $\{0, 1, 5\}$
(C) $\{0, 1, 3\}$
(D) $\{0, 1, 2\}$

14. The line, which passes through the point (4, 6) with a slope of 2, has which of the following for its equation?

(A) $y = 2x + 2$
(B) $y = 2x - 2$
(C) $y = 2x$
(D) $y = ½x - 2$

15. For what value(s) of x is the domain restricted for the function $f(x) = \frac{x^2 - y^2}{x - y}$?

(A) none
(B) $x \neq 0$
(C) $x \neq y$
(D) $x \neq -y$

WARM-UP ANSWERS

1. A There are ${}_8C_2 = \dfrac{8 \cdot 7}{2 \cdot 1} = 28$ possible combinations from list I, and ${}_7C_3 = \dfrac{7 \cdot 6 \cdot 5}{3 \cdot 2 \cdot 1} = 35$ possible combinations from list II. Altogether there are $28 \cdot 35 = 980$ or approximately 1,000 possible choices.

2. B The truth table for a conditional statement shows that when the antecedent is false but the consequent is true, the conditional is considered true.

3. C $f(x - 2) = (x - 2)^2 + 3(x - 2) - 1 = (x^2 - 4x + 4) + 3x - 6 - 1 = x^2 - x - 3$.

4. B Addition $(a + b = b + a)$ and multiplication $(ab = ba)$ are both commutative.

   Subtraction $(b - a \neq a - b)$ and division $(b \div a \neq a \div b)$ are not commutative.

5. C $\sqrt{-8} \cdot \sqrt{-2} = i\sqrt{8} \cdot i\sqrt{2} = i^2 \cdot \sqrt{8 \cdot 2} = -1 \cdot \sqrt{16} = -4$

6 C Choices A and D are linear equations and the graph of a linear equation is a straight line. Choice B is a parabola. Choice C is the only possible answer.

7. B A X B is the symbol which asks for the Cartesian Product of sets A and B. In general, if Set A has x elements and Set B has y elements, A X B has xy elements.

8. D $132_5 = 1 \cdot 5^2 + 3 \cdot 5^1 + 2 \cdot 5^0 = 25 + 15 + 2 = 42$

9. B After the first set is mailed, there are $2^1 = 2$ letters. After the second set, there are $2^2 = 4$ more. After the third, there are $2^3 = 8$ more. After the fourth, there are $2^4 = 16$ more. $16 + 8 + 4 + 2 = 30$ letters at \$0.32 each = \$9.60.

10. B The mean, or arithmetic average is $\dfrac{62 + 84 + 99 + 77 + 88}{5} = 82$.

    The median is the middle score of the numbers arranged in order 62, 77, 84, 88, 99. since there are five numbers, the middle score is the third number, or 84. The average of 82 and 84 is 83.

11. A $\sqrt{\sqrt{a^{12}}} = \sqrt{a^6} = a^3$. Note: the square root of a variable with an even exponent is the variable with the exponent divided by 2. For example $\sqrt{a^{12}} = a^6$, and $\sqrt{a^6} = a^3$.

12. A Since each dial has 6 possible numbers to choose, there will be $6 \cdot 6 \cdot 6$ possible combinations of lock numbers.

13. A The domain of the inverse is the same as the range of the given function. For $y = 2x + 1$, the range would be $2(0) + 1 = 1$, $2(1) + 1 = 3$, and $2(2) + 1 = 5$ or $\{1, 3, 5\}$.

14. B The general equation of a straight line is $y = mx + b$, where m = slope and b = y-intercept. Since we know the slope = 2, our equation is in the form $y = 2x + b$. Since the line passes through the point (4, 6), the values 4 and 6 must satisfy the equation. So, $6 = 2(4) + b$ and b, therefore, is -2. The equation is $y = 2x - 2$.

15. C If $x = y$, then the denominator is equal to zero and division by zero is impossible.

CLEP PRACTICE EXAMINATION - MATHEMATICS

PART A (45 MINUTES)

1. The product of $(x + y)(x - y) =$

   (A) $x^2 + 2xy + y^2$
   (B) $x^2 - y^2$
   (C) $x^2 + 2xy - y^2$
   (D) $x^2 - 2xy - y^2$

2. Colleen has a deck of cards and a single die. If she picks a card and rolls the die, what is the probability that the card is a spade and the die shows an even number?

   (A) $\frac{3}{4}$
   (B) $\frac{1}{8}$
   (C) $\frac{1}{4}$
   (D) $\frac{1}{2}$

3. The $\sqrt{28}$ is equal to

   (A) 5
   (B) $7\sqrt{4}$
   (C) $4\sqrt{7}$
   (D) $2\sqrt{7}$

4. In the Venn diagram, which relationship does the shaded region represent?

   (A) $(A \cap B) \cup C$
   (B) $A \cap (B \cup C)$
   (C) $A \cup (B \cup C)$
   (D) $(A \cup B) \cap C$

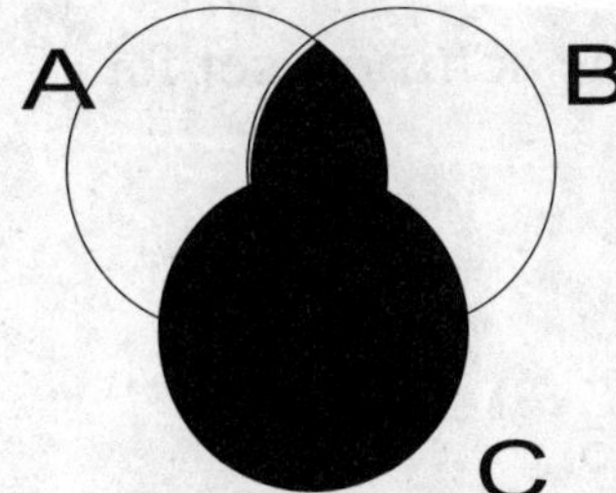

5.If $2x - 5 = 7$, then $x =$

(A) 2
(B) 12
(C) 6
(D) 1

6. What was the original price of a watch that sold at a 25% discount for $45.00?

(A) $60.00
(B) $11.25
(C) $56.25
(D) $75.00

7. Which of the following is not a factor of $x^4 - 81$?

(A) $(x^2 + 9)$
(B) $(x - 3)$
(C) $(x + 3)$
(D) $(x + 9)$

8. Sandi's scores on her math exams this semester are 93, 82, 87, 78, and 90. Her median score was:

(A) 93
(B) 86
(C) 87
(D) 78

9. If $f(X) = \frac{3X + 5}{6X}$, what is f(5)?

(A) 2/3
(B) 20
(C) 2
(D) 4

10. What is the solution set for $\left|\frac{2}{5}x - 3\right| = 11$?

(A) {35}
(B) {20}
(C) {20, 35}
(D) {-20, 35}

11. Which of the following is irrational?

(A) $\sqrt{81}$
(B) $\sqrt{7}$
(C) $\frac{\sqrt{49}}{16}$
(D) $\frac{3}{\sqrt{36}}$

12. How many prime numbers are there between 35 and 50?

(A) 3
(B) 4
(C) 5
(D) 6

13. -54 - (-28) =

(A) -26
(B) -82
(C) 26
(D) 82

14. If f(X) = (X^2 + 4), and g(Y) = (Y-5), what is f[g(4)]?

(A) 85
(B) 18
(C) 5
(D) 77

15. If $f(X) = X^2 - X + 7$, then f(-6) =

(A) 37
(B) -23
(C) -35
(D) 49

16. The conditional statement "If Marissa isn't tired, she can stay up another hour" is logically equivalent to which of the following?

(A) If she isn't tired, Marissa can't stay up another hour.
(B) If Marissa is tired, she can't stay up another hour.
(C) If she can stay up another hour, Marissa isn't tired.
(D) If she can't stay up another hour, Marissa is tired.

17. The sum of ($3x^2$ – 7x +9) and ($5x^2$ + 12x –13) is

(A) $8x^2$ + 5x - 4
(B) $8x^2$ + 19x - 4
(C) $8x^2$ - 5x - 4
(D) $8x^2$ + 5x + 22

18. $\frac{3}{\sqrt{12}}$ is equal to

(A) $\frac{\sqrt{3}}{2}$

(B) $\frac{3}{2}$

(C) $\frac{\sqrt{3}}{6}$

(D) $\frac{3\sqrt{3}}{2}$

19. Which product is larger, (-7) (5), (-7) (-5), or (7) (-5)?

(A) (-7) (5)
(B) (-7) (-5)
(C) (7) (-5)
(D) they're all equal

20. The product of (X - 8) and (X + 5) is

(A) $X^2 + 13X - 40$

(B) $X^2 + 3X - 40$

(C) $X^2 - 3X + 40$

(D) $X^2 - 3X - 40$

21. What are the factors of $8X^5 - 12X^{25}$?

(A) $4X^5(2 - 3X^5)$

(B) $8X^5(X - 4X^5)$

(C) $4X^5(2 - 3X^{20})$

(D) $8X^5(1 - 4X^{20})$

22. What is the product of 3 consecutive integers whose sum is 27?

(A) 729
(B) 720
(C) 990
(D) 504

23. If X represents an even number and Y represents an odd number, which of the following is(are) not odd?

I. X + Y + 10
II. 3Y + 5
III. XY + X

(A) I only
(B) II only
(C) I & II
(D) II & III

24. Find the product of 3^4 and 9^3.

(A) 3^7
(B) 3^{12}
(C) 3^{10}
(D) 9^7

25. Simplify $\sqrt{75}$.

(A) $5\sqrt{3}$
(B) $2\sqrt{3}$
(C) $3\sqrt{2}$
(D) $3\sqrt{5}$

26. What is the solution set for $X^2 - 11X = -24$?

(A) {6,4}
(B) {-3, -8}
(C) {3, 8}
(D) {-12, 2}

Use the graph below to solve problems 27 through 29.

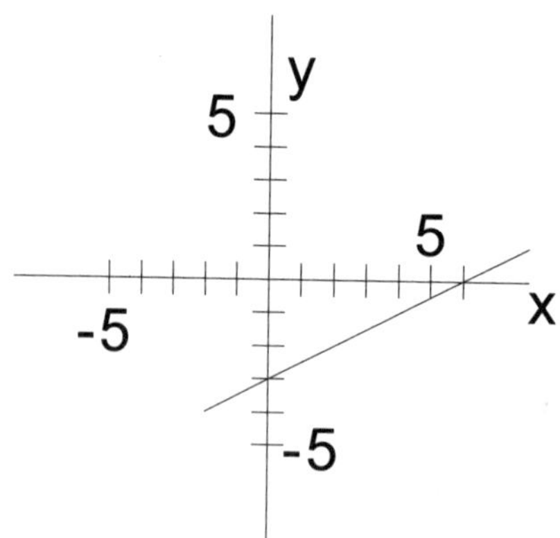

27. What is the y-intercept of the line that passes through the given points?

(A) (0, -3)
(B) (6, 0)
(C) (-6, 0)
(D) (0, 3)

28. What is the slope of the line that passes through the given points?

(A) 2
(B) 1/2
(C) -2
(D) -1/2

29. What is the equation of the line that passes through the given points?

(A) y = 1/2 X + 3
(B) y = 1/2X - 3
(C) y = -3X + 1/2
(D) y = 2X - 3

30. If A = {positive even integers < 25} and B = {multiples of 5 < 30}, then how many elements are there in A ∩ B?

(A) 3
(B) 17
(C) 2
(D) 5

31. In a recent poll of moviegoers, 7 out of 10 women said they would recommend the movie to a friend, but only 4 out of 10 men agreed. Based on this information, what is the probability that the person is a man, given that the person would not recommend the movie to a friend?

(A) 2/3
(B) 1/3
(C) 6/10
(D) 4/11

32. Which is larger, |(-3) (8)|, (-2) (-12), or (4) (6)?

(A) |(-3) (8)|
(B) (-2) (-12)
(C) (4) (6)
(D) They're all equal

PART B (45 MINUTES)

33. If A = {1,2,3,4}, B = {2,3,5,6}, what is $A \cup B$?

(A) {2,3}
(B) {1,2,5,6}
(C) {1,2,3,4,5,6}
(D) {2}

34. Which of the following is not ALWAYS true?

I. $\sqrt{a} \times \sqrt{b} = \sqrt{ab}$
II. $\sqrt{a} \div \sqrt{b} = \sqrt{a \div b}$
III. $\sqrt{a} + \sqrt{b} = \sqrt{a + b}$

(A) I only
(B) I, II only
(C) III only
(D) II only

35. Which of the following could be the equation of the graph shown?

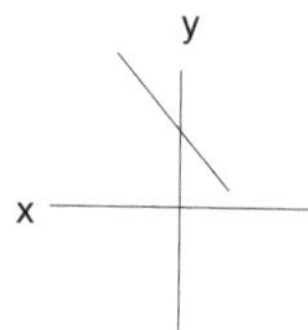

(A) y = 2x - 1
(B) y = -2x - 1
(C) y + 2x = 1
(D) y -2x = 1

36. If $f(x) = 2x^2$ and g(y) = y + 3, f[g(-2)] =

(A) -2
(B) 2
(C) -8
(D) 1

37. If p represents the statement “It is cold”, and q represents the statement “I will buy a coat”, which of the following represents the statement “If it is cold, then I will buy a coat”?

(A) $\sim p \rightarrow \sim q$
(B) $p \rightarrow \sim q$
(C) $p \rightarrow q$
(D) $q \rightarrow p$

38. What is the solution set for $|x - 2| = 3$?

(A) {5, -1}
(B) {5}
(C) {1, 5}
(D) {-5, 1}

39. How many subsets of the set {a, b, c, d, e} are there?

(A) 5
(B) 10
(C) 25
(D) 32

40. If you draw a card and replace it, and then draw another card, what is the probability that both cards are spades?

(A) 3/13
(B) 1/16
(C) 1/169
(D) 1/2

41. If $a*b = a^2+b^2$, what is $(2*1)*3$?

(A) 34
(B) 18
(C) 13
(D) 15

42. If $f(x) = \sqrt{25 - x^2}$ is to be a real number, the domain of x must be:

(A) $x > 5$
(B) $x < 5$
(C) $-5 \leq x \leq 5$
(D) $-5 \geq x \geq 5$

43. Which of the following sets is not closed under multiplication?

(A) {Even numbers}
(B) {Positive numbers}
(C) {Negative numbers}
(D) {Integers}

44. How many prime numbers are there between 50 and 60?

(A) 1
(B) 2
(C) 3
(D) 4

45. If A = {2,3,4}, B = {1,2,3}, and C = {1,5,8}, then $(A \cap B) \cup C$ =

(A) {1,2,3,4,5,8}
(B) {2,3}
(C) ∅ or { }
(D) {1,2,3,5,8}

46. The number 4 is equivalent to which of the following base 2 numbers?

(A) 100_2
(B) 11_2
(C) 10_2
(D) 101_2

47. If $a = b^3$, when b is doubled then a is

(A) tripled
(B) doubled
(C) multiplied by 8
(D) multiplied by 6

48. If a is an odd integer, which of the following must be even?

(A) 3a
(B) 3a + 2
(C) 2a - 1
(D) 3a + 1

49. If $x^2 - 3x = 28$ then x =

(A) {7, -4}
(B) {4, 7}
(C) {-7, 4}
(D) {-7, -4}

50. How many different groups of 3 people can be formed from a group of 10?

(A) 100
(B) 120
(C) 720
(D) 220

51. How many common solutions do $f(x) = x^2$ and $f(x) = 2x + 3$ have in common?

(A) 1
(B) 2
(C) 0
(D) 3

52. If $A = \{x: x \geq 1\}$, and $B = \{x: x \leq 1\}$, then $A \cap B =$

(A) $\varnothing$ or { }
(B) {1}
(C) {-1}
(D) { All Integers}

53. What is the next number in the chart below?

x	0	1	2	3	4
y	0	0	2	6	?

(A) 8
(B) 12
(C) 16
(D) 12

54. If $\sqrt{ab} = 6$, and a and b are integers, which of the following could **NOT** be the value of a – b?

(A) 0
(B) -35
(C) -16
(D) 12

55. If the conditional statement "If a = 2, then b = 3" is true, which of the following is also true?

(A) If $a \neq 2$, then $b \neq 3$
(B) If $b = 3$, then $a = 2$
(C) If $b \neq 3$, then $a \neq 2$
(D) If $b = 3$, then $a \neq 2$

56. The shaded part of the Venn diagram shown below represents which of the following?

(A) $A \cup B$
(B) $(A \cap B) \cap C$
(C) $A \cap B$
(D) $(A \cap B) \cup (A \cap C)$

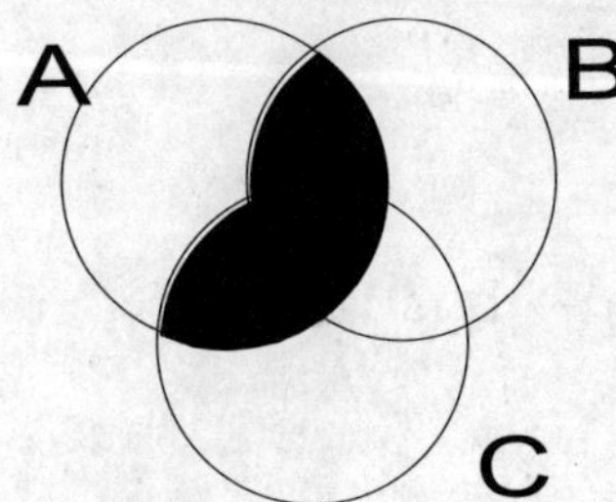

57. In the formula $P = 2\pi \frac{\sqrt{L}}{g}$, if g is a constant, what must L be multiplied by so P will be tripled?

(A) 8
(B) 3
(C) 4
(D) 9

58. Which of the following properties of logarithms are true?

I. Log AB = Log A + Log B
II. Log (A ÷ B) = Log A - Log B
III. Log A^b = b Log A

(A) I only
(B) I and III
(C) I and II
(D) I, II, and III

59. Which of the following is always true for real numbers (x and y > 0)?

I. $(x+y)^a = x^a + y^a$
II. $(x^a)(y^a) = (xy)^a$
III. $x^a \div y^a = (x \div y)^a$

(A) I and III
(B) II and III
(C) I, II, and III
(D) None

60. If x is a real number, $x^2 < x$ is true for:

(A) $x > 1$
(B) $x < 1$
(C) $x = 0$
(D) $0 < x < 1$

61. The statement $\sim a \rightarrow b$ is false if:

(A) a is true
(B) b is true
(C) b is false and a is true
(D) a is false and b is false

62. If $\sqrt{-1}$ is represented by i, what is (4 - 2i)(4 + 2i)?

(A) 20
(B) 16
(C) 16 - 4i
(D) 20i

63. The graph of the straight line y = f(x) passes through the point (3, 5) and also through (4, a). If the slope of the line is 3, what is the value of a?

(A) 6
(B) 12
(C) 14
(D) 8

64. A box contains cards numbered 1 through 21. What is the probability that the first card drawn is divisible by 2 and the second card drawn is divisible by 5 if the first card is replaced after drawing?

(A) $\frac{40}{441}$
(B) $\frac{14}{21}$
(C) $\frac{12}{21}$
(D) $\frac{44}{441}$

65. If $\log_{10} 5 = b$ and $\log_{10} 7 = c$, then $\log_{10} 35 = ?$

(A) b + c
(B) b - c
(C) bc
(D) $\frac{b}{c}$

ANSWERS TO PRACTICE EXAMINATION

PART A

1. B x multiplies both x and (-y), and (+y) multiplies both x and (-y). The result is $x^2 + xy - xy - y^2$. The middle terms cancel each other out, and the answer is $x^2 - y^2$.

2. B These are independent events, so the answer is the product of the individual probabilities.

$$P(\text{spade}) = \frac{1}{4}$$

$$P(\text{even}) = \frac{1}{2}$$

$$P(\text{spade and even}) = \frac{1}{4} \cdot \frac{1}{2} = \frac{1}{8}$$

3. D $\sqrt{28} = \sqrt{4}\sqrt{7} = 2\sqrt{7}$

4. A The solution is the area that is in both circles A and B plus the area enclosed by circle C.

5. C

$$\begin{aligned} 2x - 5 &= 7 \\ +5 &= 5 \\ \hline 2x + 0 &= 12 \end{aligned}$$

$$x = \frac{12}{2} = 6$$

6. A If the discount is 25%, then \$45.00 represents 75% of the original cost.

$.75x = 45$

$x = 45/.75 = 60$

7. D $x^4 - 81$ is the difference between 2 squares. Initially, we write it as the product of $(x^2 - 9)(x^2 + 9)$. $(x^2 - 9)$ is also the difference of 2 squares whose factors are $(x - 3)(x + 3)$. The 3 factors are $(x - 3)(x + 3)(x^2 + 9)$. The term $(x + 9)$ is not a factor.

8. C The median score is the number for which there are as many scores below it as above it. If you put the scores in order, you can see that the middle score is 87. 86 is the mean.

9. A $\frac{3(5)+5}{6(5)} = \frac{15+5}{30} = \frac{20}{30} = \frac{2}{3}$

10. D 2/5 x - 3 = 11 and 2/5 x - 3 = -11

2/5 x = 14 and 2/5 x = -8

x = 35 and x = -20

11. B $\sqrt{81} = 9$, $\frac{\sqrt{49}}{16} = \frac{7}{16}$, and $\frac{3}{\sqrt{36}} = \frac{3}{6} = \frac{1}{2}$. $\sqrt{7}$ is the only one that cannot be expressed as a fraction of integers.

12. B The prime numbers are 37, 41, 43, and 47.

13. A -54 - (-28) = -54 + 28 = -26

14. C g(4) = 4 - 5 = -1. f (-1) = $[(-1)^2 + 4]$ = 1 + 4 = 5

15. D f (-6) = $(-6)^2$ - (-6) + 7

= 36 + 6 + 7

= 49

16. D If A = tired and B = stay up, then the statement can be expressed as ~A → B. The contrapositive is logically equivalent, which in this case would be ~B → A, or choice D.

17. A Add like terms together, combining the coefficients using our rules for signed numbers.

18. A Multiply the numerator and denominator by $\sqrt{3}$. The result is $\frac{3\sqrt{3}}{\sqrt{36}} = \frac{3\sqrt{3}}{6} = \frac{\sqrt{3}}{2}$.

19. B The first and third products are negative, the second product is positive.

20. D x multiplies both x and (+5), and (-8) multiplies both x and (+5). The result is $x^2 + 5x - 8x - 40$. Combining the middle terms, we get $x^2 - 3x - 40$.

21. C 4 goes evenly into both 8 and 12. x^5 goes evenly into both x^5 and x^{25}. The common factor is $4x^5$. Remember, when you divide x^5 into x^{25}, you subtract the exponents.

22. B The easiest way to solve this problem is to divide 27 by 3, which equals 9, and add and subtract 1 from 9. The three numbers are 8, 9, and 10. Their product is 720.

23. D Pick any even number for X, say 2, and any odd number for Y, say 1. Then:

I = 2 - 1 + 10 = 13
II = 3(1) + 5 = 8
III = (2)(1) + 2 = 4.

II and III are not odd.

24. C Before you combine the exponents, you must have a common base, so you must rewrite 9^3: $9^3 = (3^2)^3 = 3^{2x3} = 3^6$. Then just add the exponents of 3 together.

25. A $\sqrt{75} = \sqrt{25}\sqrt{3} = 5\sqrt{3}$.

26. C Add 24 to both sides so that the equation is equal to 0. Then write in factored form (x - 8) (x - 3) = 0. x - 8 = 0 or x - 3 = 0. Solving for x, x = 8 or 3.

27. A The y-intercept is the point where the x value is 0, so the only possible answers are (A) or (D). Since the line crosses the y-axis below the x-axis, we know that the y-intercept is negative.

28. B The line is rising from left to right, so the slope is positive. The slope is a slow rise, so it's probably a fraction less than 1. To calculate the slope, we take the difference in the y coordinates over the difference in the x coordinates.

$$\frac{(-3-0)}{(0-6)} = \frac{-3}{-6} = \frac{1}{2}$$

29. B y = mx + b; m = $\frac{1}{2}$ and b = -3.

30. C The intersection is {10, 20}.

31. A There are 6 men who didn't like the movie and 3 women, for a total of 9 people. The probability that the person was a man is $\frac{6}{9}$ or $\frac{2}{3}$.

32. D In each case the result is a positive 24.

PART B

33. C The union contains any element that is in either set.

$A \cup B = \{1, 2, 3, 4, 5, 6\}$

34. C Property III is generally not true. Take, for example, a = 4 and b = 9. Then $\sqrt{4}+\sqrt{9}=2+3=5$, which is not equal to $\sqrt{4+9}=\sqrt{13}$.

35. C We can see by inspection that the line has a negative slope and a positive y-intercept. Writing answer C in the form y = mx + b, we get y = -2x + 1. It is the only response that has a negative slope and a positive y-intercept.

36. B g(-2) = -2 + 3 = 1. $F(1) = 2(1)^2 = 2$.

37. C The "if-then" statement is called the conditional and its logic symbol is $\rightarrow$. So, "if p, then q" becomes $p \rightarrow q$.

38 A |x - 2| is true if either x - 2 = 3 or x - 2 = -3. x - 2 = 3 when x = 5 and x - 2 = -3 when x = -1.

39. D The number of subsets of a set with n elements is given by 2^n. Since this set has 5 elements, there are 2^n = 32 subsets. (Note: this includes the null set and the set itself.)

40. B The probability that two independent events will happen is found by multiplying their individual probabilities together:

P(1st spade) = $\frac{1}{4}$; P(2nd spade) = $\frac{1}{4}$; P(drawing 2 spades) = $\frac{1}{4} \cdot \frac{1}{4} = \frac{1}{16}$.

41. A If $a * b = a^2 + b^2$, then

$$\begin{aligned}(2*1)*3 &= (2^2+1^2)*3\\ &= (4+1)*3\\ &= 5*3\\ &= 5^2+3^2\\ &= 25+9\\ &= 34\end{aligned}$$

42. C If f(x) is to be a real number, the expression under the radical sign must be zero or greater since the square root of a negative number is not a real number. Therefore, x must be less than or equal to 5 and greater than or equal to -5.

43. C The product of two even numbers is even. This set is closed under multiplication. The product of two positive numbers is positive, so this set is also closed. The product of two negative numbers is positive. This set, therefore, is not closed under multiplication.

44. B 53 and 59 are the only prime numbers between 50 and 60.

45. D $A \cap B = \{2, 3\}$. The union of this with the set $\{1, 5, 8\} = \{1, 2, 3, 5, 8\}$

46. A $2^0 = 1$, $2^1 = 2$, $2^2 = 4$. Therefore, $4_{10} = 100_2$.

47. C If $a = b^3$, then substitute twice b or 2b for b = $(2b)^3 = 8b^3$. Since $a = b^3$, this = 8a.

48. D If a is odd, 3a is also odd, because the product of two odd numbers is odd. Adding 1 to an odd number is the sum of two odd numbers, which is even.

49. A

$$x^2 - 3x = 28$$
$$x^2 - 3x - 28 = 0$$

thus, $(x - 7)(x + 4) = 0$

$$x - 7 = 0 \qquad x + 4 = 0$$
$$x = 7 \qquad x = -4$$

50. B The number of committees would be given by: ${}_{10}C_3 = \frac{10 \cdot 9 \cdot 8}{3 \cdot 2 \cdot 1} = 120$.

51. B The common solutions are the points where the graphs intersect. The graphs below intersect at two points.

NOTE: The algebraic solution can be found be setting the equations equal to each other:

$$X^2 = 2x + 3$$
$$X^2 - 2x - 3 = 0$$
$$(x - 3)(x + 1) = 0$$
$$x = 3, \quad x = -1$$

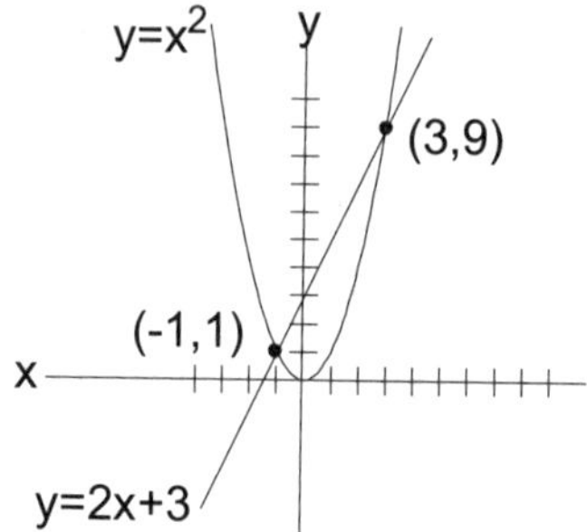

Substituting each x and solving for y: $y = 3^2 = 9$ and $y = (-1)^2 = 1$.

The common points are (3, 9) and (-1, 1).

52. B A is the set of numbers greater than or equal to 1. B is the set of numbers less than or equal to 1. The only common element is 1.

53. B By squaring each x and subtracting x from the total you get y. Therefore, the value of y for x = 4 is, $4^2 - 4 = 16 - 4 = 12$.

54. D If $\sqrt{ab} = 6$, then by squaring both sides, ab = 36. The factors of 36 are: 36 x 1, 18 x 2, 12 x 3, 9 x 4, and 6 x 6. Only 12 can not be found by subtracting a and b.

55. C If $a \neq 2$, it does not necessarily follow that b cannot equal 3. For choice B it is possible that b = 3 under other conditions besides when a = 2. However, it is true that if $b \neq 3$ then a cannot equal 2, because if a = 2, then b = 3.

56. D The shaded portion of the Venn diagram represents the intersection (or common part) of A and B together with the intersection of A and C. In symbols, this is $(A \cap B) \cup (A \cap C)$.

57. D Since L is under the radical or square root symbol, and g is a constant, to extract a quantity three times larger the quantity under the radical would have to be nine times larger. (The $\sqrt{9} = 3$.)

58. D Since logarithms are really exponents, all three properties are true. The first corresponds to adding exponents in multiplication. The second corresponds to subtracting exponents in division. The third corresponds to raising an exponent to a power.

59. B Only statement I is false. By substituting numbers you see that $(3 + 4)^2$ does not equal $3^2 + 4^2$. Also, you know that (a + b)(a + b) is equal to $a^2 + 2ab + b^2$, which is not equal to $a^2 + b^2$.

60. D The square of a number is less than the number itself when the number is a positive fraction less than 1.

61. D The logic symbols translate, "If not a, then b." For this statement to be false, both the antecedent (a) and the consequent must be false because if b is false, then a is true.

62. A $(4 - 2i)(4 + 2i) = 4^2 - (2i)^2 = 16 - 4i^2$. Since $i = \sqrt{-1}$, then $i^2 = \sqrt{-1} \cdot \sqrt{-1} = -1$. Therefore $16 - 4i^2 = 16 - 4(-1) = 16 + 4 = 20$.

63. D The slope of a straight line can be found by dividing the difference in the y coordinates by the difference in the x coordinates. In this problem, that would be equal to: $\frac{(a-5)}{(4-3)} = 3$. Solving for a, we get a = 8.

64. A There are 10 favorable numbers divisible by 2: 2, 4, 6, 8, 10, 12, 14, 16, 18, 20. The probability is $\frac{10}{21}$. There are only 4 favorable numbers that are divisible by 5: 5, 10, 15, 20. The probability is $\frac{4}{21}$. The probability of both is $\frac{10}{21} \times \frac{4}{21} = \frac{40}{441}$.

65. A The log of the product of two numbers is equal to the sum of the logs. Since $35 = 7 \cdot 5$, log 35 = log 7 + log 5 = b + c.